FLAT PANEL DISPLAYS
Advanced Organic Materials

RSC Materials Monographs

Series Editor: J.A. Connor, *Department of Chemistry, University of Kent, Canterbury, UK*

Advisory Panel: G.C. Allen (*Bristol, UK*), D.J. Cole-Hamilton (*St Andrews, UK*), W.J. Feast (*Durham, UK*), P. Hodge (*Manchester, UK*), M. Ichikawa (*Sapporo, Japan*), B.F.G. Johnson (*Cambridge, UK*), G.A. Ozin (*Toronto, Canada*), W.S. Rees (*Georgia, USA*)

The chemistry of materials will be the central theme of this Series which aims to assist graduates and others in the course of their work. The coverage will be wide-ranging, encompassing both established and new, developing areas. Although focusing on the chemistry of materials, the monographs will not be restricted to this aspect alone.

Polymer Electrolytes
by Fiona M. Gray, *School of Chemistry, University of St. Andrews, UK*

Flat Panel Displays: Advanced Organic Materials
by S.M. Kelly, *Department of Chemistry, University of Hull, UK*

How to obtain future titles on publication

A standing order plan is available for this series. A standing order will bring delivery of each new volume immediately upon publication. For further information, please write to:

Sales and Customer Care
Royal Society of Chemistry
Thomas Graham House
Science Park
Milton Road
Cambridge
CB4 0WF
Telephone: +44(0) 1223 420066

RSC
MATERIALS
MONOGRAPHS

Flat Panel Displays
Advanced Organic Materials

S.M. Kelly
Department of Chemistry, University of Hull, UK

RS•C
ROYAL SOCIETY OF CHEMISTRY

ISBN 0-85404-567-8

A catalogue record for this book is available from the British Library.

Published by The Royal Society of Chemistry,
Thomas Graham House, Science Park, Milton Road
Cambridge CB4 0WF, UK

For further information see our web site at www.rsc.org

Typeset by Paston PrePress Ltd, Beccles, Suffolk
Printed by MPG Books Ltd, Bodmin, Cornwall

Preface

Liquid crystals have found wide commercial application over the last 25–30 years in electro-optical flat panel displays (FPDs) for consumer audio-visual and office equipment such as watches, clocks, stereos, calculators, portable telephones, personal organisers, note books and laptop computers. There are many other applications for liquid crystal displays (LCDs) such as information displays in technical instruments and in vehicles initially as clocks, then speedometers to a lesser extent and now increasingly as navigation and positional aids or entertainment consoles. They are also used in low-volume niche products such as spatial light modulators and generally as very fast light shutters. More importantly they have come to dominate the displays market in portable instruments due to their slim shape, low-weight, low voltage operation and low power consumption, see Table 1.1. LCDs are now starting to win market share from cathode ray tubes (CRTs) in the computer monitor market. The market share of LCDs of the total market for displays is expected to significantly increase over the next decade. There are a number of existing competing flat-panel display technologies, such as plasma display panels (PDPs), vacuum fluorescence displays (VFDs), inorganic semiconductor light-emitting diodes (LEDs), digital micromirror devices (DMDs) and field emission displays (FEDs). However, these have relatively small shares of the overall displays market, see Table 1.1. The most promising technology for FPDs being developed at the moment is represented by organic light-emitting diodes (OLEDs) using either low-molar-mass (LMM) materials (small molecules) or light-emitting polymers (LEPs). The first production lines for LEP technology have recently been commissioned and products are commercially available. However, in spite of these competing FPDs, especially OLEDs which are expected to exhibit significant growth, the value of LCDs is still expected to exceed that of CRTs in the near future. Manufacturing facilities for flat-panel displays are very capital intensive, *e.g.* a plant for TN (twisted nematic)-LCDs with active matrix addressing can cost upwards of $1 billion. As a consequence of a combination of factors, such as the capital already invested in LCD plants, which have to be depreciated, the steadily decreasing unit cost of LCDs and the expanding market requirement for them in existing products, it will take many years for competing technologies to gain a significant market share in the displays market in general. In particular, LCDs can be expected to maintain a dominant position in the market for portable displays.

The successful development of LCD technology was dependent on parallel developments and progress in an unusual combination of scientific disciplines such as synthetic organic chemistry, physics, electronics and device engineering. These include improvements in batteries, polarisers, electrodes, CMOS drivers, spacers, alignment layers and nematic liquid crystals. These developments were made in response to a clear market requirement for a low-voltage, low-power-consuming flat-panel display screen for portable, battery-operated instruments in order to display graphic and digital information of ever more increasing volume, speed and complexity. In this monograph we will attempt to illustrate this development using the most important types of LCD currently in large-scale manufacture. We will not describe LCDs using smectic liquid crystals, although they may have the potential to become a major commercial product, since they have failed to make a commercial breakthrough after more than two decades of research and development. All of these types of LCDs can be modified to use nematic gels or polymer-dispersed liquid crystals. Therefore, these will not be dealt with here in any detail as the general principles of operation and electro-optical effects are essentially the same.

OLEDs using electroluminescent small molecules have been in continuous development for over 30 years and intensive development for at least the last decade. Only now are OLEDs using low-molar-mass materials being manufactured on any significant scale. The development of OLEDs using conjugated organic polymers has been able to profit from the know-how and technology developed for low-molar-mass OLEDs. However, the time from discovery to manufacture has been much shorter, *i.e.* about 10 years. The first pilot plants for OLEDs using LEPs have just been commissioned at Philips and UNIAX and commercial products incorporating LEPs from Dow Chemicals, Clariant or Covion are to expected on the market in 2001 based on technology under licence, at least in part, from CDT.

The development and successful commercialisation of LCDs and OLEDs as flat panel displays in consumer and industrial products and instruments is illustrative of the general dependence of consistent improvement in the performance of flat panel displays as a consequence of the invention, laboratory preparation, evaluation, optimisation, scale-up and then large-scale manufacture of organic compounds and their mixtures in high purity and at acceptable cost with the required spectrum of physical properties. This development in materials chemistry will be described in detail in the following chapters. The interdependence of technologies in electro-optics, electronics and organic chemistry is illustrated by the imaginative use of liquid crystals in OLEDs as charge-carrier transport layers and as electroluminescent materials as the source of plane polarised light for hybrid OLED-LCDs with intrinsically higher brightness.

The fundamental electro-optical principles of LCDs and OLEDs with their relative advantages and disadvantages for a diverse range of specific applications are described in this monograph. These specifications then prescribe the relative and absolute magnitude of a spectrum of physical properties for nematic liquid crystals and electroluminescent organic materials to fulfil in order for these types of flat panel displays to function as efficiently as possible and attain their maximum potential for a particular application. The nature of the nematic liquid crystalline state and the origin of electroluminescence in small organic molecules and organic conjugated polymers is described sufficiently to understand the correlation between molecular structure and those physical properties of relevance to LCDs and OLEDs, at least where such correlations are understood. This monograph concentrates primarily on the developments in the design and synthesis of these two different classes of organic materials specifically for use in the two most important types of FPDs using organic compounds. Therefore, the theoretical, especially mathematical, background to the phenomenon of liquid crystallinity and organic electroluminescence are kept to a necessary minimum.

Contents

Abbreviations, Acronyms and Symbols

AA	Active Addressing
AC	Alternating Current
Alq_3	Aluminium *tris*(2-hydroxyquinolate)
AM	Active Matrix
C (F)	capacitance (farads)
CCH-5	*trans*-1-(*trans*-4-Cyanocyclohexyl)-4-pentylcyclohexane
CDT	Cambridge Display Technology
Ch	Cholestolic
CNPC	Cholesteric Nematic Phase Change
Cr	Crystal
CRT	Cathode Ray Tube
d (μm)	cell gap (micrometers)
DAP	Deformation of Vertically Aligned
DC	Direct Current
DMD	Digital Micromirror Devices
DSM	Dynamic Scattering Mode
DSTN	Double Super Twisted Nematic
E_{abs} (eV)	energy of absorption (electron-volts)
E_{em} (eV)	energy of emission (electron-volts)
ECB	Electrically Controlled Birefringence
EL	Electroluminescence
E_p (eV)	phonon energy (electron-volts)
ETL	Electron-Transport Layer
FDP	Flat Panel Display
FED	Field Emission Display
g	Kirkwood Foehlich factor
GaAsP	Gallium Arsenide Phosphorous
GH	Guest–Host
h (J s)	Planck's constant (Joule seconds)
HN	Homeotropic Nematic
HOMO	Highest Occupied Molecular Orbital

HTL	Hole-Transport Layer
I	Isotropic
$I \, (\text{mA cm}^{-2})$	current density (milliamp per centimetre squared)
IPS	In-Plane Switching
ITO	Indium–Tin Oxide
I-V	Current-Voltage
k_{11}, k_{22}, k_{33} (N)	elastic constants (newton)
$L \, (\text{cd m}^{-2})$	brightness (candelas per meter squared)
LCD	Liquid Crystal Display
LED	Light Emitting Diode
LEP	Light Emitting Polymer
LMMM	Low Molar Mass
LUMO	Lowest Unoccupied Molecular Orbital
MBBA	N-(4-methoxybenzylidene)-4'-butylaniline
MEH-PPV	Poly[2-methoxy-5-(2-ethylhexyloxy)]-4-phenylene vinylene
N	Nematic
n	average refractive index
N*	Chiral Nematic
n_e	refractive index of the extraordinary ray
n_o	refractive index of the ordinary ray
NW	normally white
OLED	Organic Light-Emitting Diode
OMI	Optical Mode Interference
$p \, (\mu\text{m})$	pitch (micrometres)
PANi	Poly(aniline)
PBD	2-(4-tert.-butylphenyl)-5-(biphenyl-4-yl)-1,3,4-oxadiazole
PCH-5	_trans_-1-(4-cyanophenyl)-4-pentylcyclohexane
PD	Polymer Dispersed
PDP	Plasma Display Panel
PECH-5	_trans_-1-[2-(4-cyanophenyl)ethyl]-4-pentylcyclohexane
PEDOT	Poly(3,4-ethylenedioxythiophene)
PET	Poly(ethylene terephthalate)
Pixel	Picture Element
PL	Photoluminescence
PMMA	Poly(methyl methacrylate)
PPP	Poly(p-phenylene)
PPV	Poly(p-phenylene vinylene)
PVC	Poly(vinylcinnamate)
PVK	Poly(N-vinylcarbazole)
RGB	Red Green Blue
RMS	Root Mean Square
S	order parameter
SAM	Self-Assembled Monolayer
SBE	Super Birefringent Effect
SCL	Space-Charge Limited
SmA	Smectic A

SmA*	Chiral Smectic A
SmB	Smectic B
SmX	Smectic X
Sol (wt%)	solubility (weight percent)
SSF	Surface Stabilised Ferroelectric
STN	Super Twisted Nematic
T (°C)	temperature (degrees centigrade)
TFT	Thin-Film Transistor
T_g (°C)	glass transition temperature (degrees centigrade)
TN	Twisted Nematic
T_{N-I} (°C)	nematic clearing point (degrees centigrade)
t_{off} (ms)	switch-off time (milliseconds)
t_{on} (ms)	switch-on time (milliseconds)
T_{red} (°C)	reduced temperature (degrees centigrade)
UV	Ultra Violet
V_{10} (V)	voltage at 10% light transmission (volts)
V_{50} (V)	voltage at 50% light transmission (volts)
V_{90} (V)	voltage at 90% light transmission (volts)
VAN	Vertically Aligned Nematic
V_c (V)	capacitive threshold voltage (volts)
VFD	Vacuum Fluorescence Display
V_{NS} (V)	non-select voltage (volts)
V_{op} (V)	operating voltage (volts)
V_S (V)	select voltage (volts)
V_{th} (V)	threshold voltage (volts)
Δn	birefringence
$\Delta\delta$ (°)	optical retardation (radians)
$\Delta\varepsilon$	dielectric anisotropy
Φ (°)	twist angle (degrees)
δ (μm)	optical path difference (micrometres)
$\varepsilon_{\parallel}$ (F m^{-1})	dielectric permittivity measured parallel to the director (faradays per metre)
$\varepsilon_{\perp}$ (F m^{-1})	dielectric permittivity measured perpendicular to the director (faradays per metre)
ϕ (%)	quantum efficiency (percent)
ϕ_F (%)	quantum efficiency of fluorescence (percent)
γ	double charge injection factor
γ_1 (mm^2 s^{-1})	rotational viscosity (millemetre squared per second)
η_E (lm W^{-1})	power efficiency (lumens per watt)
η_{ext} (%)	external quantum efficiency (percent)
η_{int} (%)	internal quantum efficiency (percent)
η_R (%)	singlet formation efficiency (percent)
η (cP)	flow viscosity (centipoise)
$\lambda_{max, abs}$ (nm)	wavelength of maximum absorption (nanometers)
$\lambda_{max, em}$ (nm)	wavelength of maximum emission (nanometers)
μ_e (cm^2 V^{-1} s^{-1})	electron mobility (centimetre squared per volt per second)

μ_h (cm^2 V^{-1} s^{-1}) hole mobility (centimetre squared per volt per second)
μ (D) dipole moment (debye)
θ (°) tilt angle (degrees)
σ (Ω cm^{-1}) resistivity (ohm per centimetre)
υ_{abs} (s^{-1}) frequency of absorption (per second)
υ_{em} (s^{-1}) frequency of emission (per second)

CHAPTER 1

Flat Panel Displays

1 Flat Panel Displays

The cathode ray tube (CRT) is still the dominant electro-optical display device today, although this is expected to change in the next few years. The CRT is still the benchmark display in terms of cost and performance. There are many areas of the market for electro-optic displays where one or more of the competing flat-panel display technologies offers a superior technological performance to a CRT, see Table 1.1.[1] Perhaps the most important are portable applications where the combination of physical properties, such as low power consumption, low operating voltage and light-weight of liquid crystal displays (LCDs) is clearly superior to that of CRTs. Most flat panel displays are emissive displays, *i.e.* they emit light without requiring absorbing polarisers like LCDs. Therefore, their brightness and viewing angle dependence are fundamentally superior to those of LCDs, which modulate the intensity of transmitted light from some independent internal or external light source. Therefore, they must be used with a back-light where insufficient ambient light is present. Light-emitting flat panel displays (FPDs) offer superior performance in poor ambient light conditions or in the dark whereas reflective FPDs are clearly superior in a bright light

Table 1.1 *Estimated world-wide market share of flat panel displays in the year 2000*[1]

Type of flat panel display (FPD)	*Number of units*
Liquid crystal displays (LCDs) with segmented characters	1 470 000 000
Super-twisted nematic liquid crystal displays (STN-LCDs)	45 000 000
Liquid crystal displays (AM-TFT-LCDs) with active matrix thin film transistor addressing	48 000 000
Organic electroluminescent displays (OLEDs)	300 000
Plasma display panels (PDPs)	630 000
Field emission displays (FEDs)	540 000
Inorganic semiconductor light-emitting diodes (LEDs)	181 000 000
Vacuum fluorescent displays (VFDs)	166 000 000
Total	1 900 000 000

1

environment. The former are not visible in the dark and the latter are washed out in bright light.

A flat panel display may be several millimetres or several centimetres thick. There are many technologies capable of being used to create a flat panel display. The most important flat panel displays are described briefly below; the two most important are LCDs and OLEDs, which are the subject of this monograph. Both require organic materials in order to function. Therefore, these are described in much more detail.

A high-information-content display must be capable of displaying an equivalent amount of information as a CRT of comparable size. The major segments of the displays market in general for CRTs are as television screens and static, *i.e.* non-portable computer monitors.

Emissive displays are intrinsically brighter than commercial LCDs currently available, even those with a strong back-light. The use of crossed, absorbing polarisers limits the maximum intensity of incident light transmitted to 25%. Therefore, a large amount of research and development effort is being devoted to optimising internal reflectors, which replace one polariser, optical retarders and different types of LCDs, which use either one polariser or no polarisers.

Advances in optimising the physical properties in organic materials such as nematic liquid crystals, electroluminescent small molecules and polymers are the topic of this monograph. Oligomers are intermediate compounds between low-molar-mass materials (small molecules) and polymers and serve as model compounds for studying polymers without the polydispersity of the latter. However, they are not used commercially, and probably will not be in the foreseeable future. Therefore, they will not form part of this monograph. Parallel developments in device peripherals such as organic polymer alignment layers, organic optical retarders and polarisers are also important. These are also described briefly. However, a satisfactory electro-optic performance of a particular display type is not always a sufficient criterium for commercialisation. The properties of other electro-optic components, such as the cost of drivers can play a decisive role in deciding whether a particular display technology is manufactured at all, occupies a niche in the displays market or is manufactured in large volumes. However, these parameters often depend on the fundamental mode of operation of a particular display technology. These are described and compared briefly in this chapter for FPDs in general and in much more detail in Chapters 2–6 for LCDs and organic light-emitting diodes (OLEDs).

Flat-Panel Cathode Ray Tubes[2]

The production of flat-panel cathode ray tubes (CRTs) is essentially a fabrication issue. The basic principle of operation is the same as a standard CRT. Electrons are emitted from a hot cathode. These are guided by a magnetic field to the glass screen coated in a layer of phosphorescent material. Upon impact the energy of the electron is transferred to the phosphor and light is emitted. A regular pattern of red, green and blue phosphors creates a dense pattern of pixels, which allows the generation of full colour. A gas plasma discharge may also be used as a source of electrons.

The high voltage requirement, *i.e.* < 200 V, and power consumption are the main restrictions to the utilisation of flat CRTs due to their incompatibility with battery operation over an extended period of time due to the high voltages and power consumption required. Other flat panel displays are more suitable and are usually preferred for portable, hand-held applications. The difficulty associated with manufacturing flat, rectangular large-area cathode ray tubes is an added problem preventing their use as screens for portable instruments. Such large CRTs would still be relatively heavy despite their relatively flat, thin construction due to the weight of the thick-walled glass vacuum tube required for mechanical stability.

Plasma Display Panels[3]

Plasma display panels (PDPs) based on an emissive gas discharge phenomenon were invented over 30 years ago. Indeed large-area plasma panel displays have been commercially available since 1970. Monochrome PDPs use visible light emitted under the action of a small electric current flowing between the electrodes. Full colour displays use UV emission at 150 nm or 173 nm to address an alternating array of red, green and blue phosphorescent strips. Short response times and steep electro-optic transmission curves facilitate the fabrication of very large-area, high-information-content plasma display panels (> 60" diagonal). However, their high cost and substantial size and weight has restricted their acceptance for the consumer market. Moreover, flat-panel plasma displays require a large number of expensive, high-voltage, alternating current (AC) or direct current (DC) drivers. Furthermore, the high operating voltages and power consumption prohibit their use in portable, battery-operated applications. Therefore, PDPs have traditionally been used for non-portable, high-cost, low-volume display applications, which are far less cost-sensitive, such as industrial, commercial or military applications. LCDs with a very large area and high information content, *e.g.* for TVs with a 40" diagonal and above, are still very expensive and not competitive with PDPs. However, the unit-cost of large-area, high-information-content PDPs is also steadily decreasing. Consequently, the acceptance of PDPs as very large televisions and monitors in the consumer market is gradually increasing. Unfortunately the large pixel size (≈ 1 mm) gives rise to relatively low resolution and a grainy appearance for short viewing distances.

Vacuum Fluorescence Displays[4]

Vacuum fluorescent displays (VFDs) are strongly related to flat-panel CRTs. Electrons are ejected from a cathode source, traverse a vacuum and then strike a pattern of triodes with individual anodes covered in red, green and blue phosphorescent material. However, the operating voltages, *e.g.* 12 V, and power consumption are much lower than those found for CRTs and PDPs.

The fabrication costs of VFDs are also relatively low. They are rugged with long operating lifetimes. Therefore, small VFDs have been manufactured in large volume for several decades for a variety of applications, *e.g.* as part of car dashboards or orientation and navigation systems.

Once again the major problems associated with the commercialisation of large VPDs is their manufacture. These include increasing weight of the glass tubes, which are necessarily thick walled. Precise spatial matching of the cathode and anode matrices is also problematical at large display size. Multiplex addressing of larger displays results in unacceptably high operating voltages, *e.g.* 100 V, for battery-operated devices. VFDs with active matrix addressing use much lower operating voltages, but are correspondingly more expensive.

Field Emission Displays[5,6]

Field emission displays (FEDs) utilise a very similar technology to the CRT tube, *i.e.* electron-impact induced light emission from a flat screen coated with alternating strips of red, green and blue organometallic phosphors. However, the main difference is that the electrons are not generated as a beam from a hot cathode, which is then directed by a magnetic field towards the screen, as in a CRT, but are emitted individually from a dense matrix of pointed pixel electrodes covering the active cathode area of the display. The narrow gap between the flat phosphor screen on top of the anode and the planar emission cathode layer and substrate is small, *e.g.* 2 mm. Therefore, considerably lower voltages are required for FEDs than for CRTs. However, the current density is significantly higher. This mode of operation allows light-weight flat panels to be constructed with a relatively low power consumption, wide viewing angle, high brightness, video-rate addressing and ruggedness. The contrast is generally relatively low (> 20:1). Flat-panel FEDs are available as monochrome and full-colour commercial products, although with a relatively small screen size (5" diagonal) for the moment. Larger prototypes have been demonstrated (12" diagonal). However, the most important factor holding back the wide-scale adoption of FEDs as a flat-panel display is the high operating voltage (> 20 V). This inhibits their use in portable device applications due to short battery life-times.

Digital Micromirror Devices[7,8]

Digital light processing devices use micro-electromechanical systems referred to as a digital micromirror device (DMD). An array of rectangular polished aluminium mirrors, *e.g.* 640×480 pixels, each individual mirror situated above a CMOS memory chip, can be addressed by an applied voltage to reflect light through a microlens in the on-state or deflect light in the off-state. This is a bistable, black-on-white memory effect compatible with video-rate addressing with high contrast (> 100:1) and high brightness (≈ 300–400 lumens). The mirrors are fabricated in a series of lithographic steps on a single substrate. Grey scale can be realised using pulsed applied voltages with full colour achieved using colour filters. Therefore, DMDs are used as high-information-content front or rear projection devices, especially for home cinema and commercial cinema or stadia applications. However, they are essentially projection devices and the size and weight of the projector and light source are too large for portable applications.

Inorganic Semiconductor Light-Emitting Diodes[9]

Light-emitting diodes (LEDs) are flat panel displays which emit light under the action of an electric current passing through the emissive layer. Electroluminescence in inorganic semiconductors was discovered before the corresponding effect in organic materials was found. Consequently the first commercial alpha numeric display devices fabricated in the early 1960s used electroluminescence inorganic semiconductor materials, such as GaAs/P or ZnS/Mn on a glass substrate sandwiched between two dielectric layers. These separate the emissive material from the electrodes and limit the amount of current flowing through the display. Pulses of alternating current result in light emission. Monochrome semiconductor inorganic LEDs are manufactured on a large scale and are found in many electronic instruments.

High-information-content LEDs using inorganic semiconductors have been produced with active matrix addressing using thin film transistors on a silicon substrate. However, the size of the displays is limited by the amount of power consumed by the large number of pixels due to the high capacitance at each individual pixel. The power consumption of a large-area LED, such as a notebook computer screen, would be considerable, *e.g.* 100 W. Other addressing problems, such as non-uniform grey scale due to the steep curve of brightness against voltage, also become disproportionately acute with increasing display size.

Organic Light-Emitting Diodes[10]

There remains an enormous potential for light-emitting diodes using organic materials (OLEDs) due to their advantageous combination of physical properties, such as ease of processing, robustness and an almost infinite possibility for modification, *e.g.* wavelength of emission, by suitable materials chemistry design and synthesis. The process of electroluminescence from organic materials is essentially the same as that from inorganic materials except that the emission takes place from a molecular excited state and not from an atomic excited state (energy level). Therefore, the bandwidth of emission is broader due to molecular vibrations. OLEDs are characterised by low operating voltages and power consumption, wide viewing angles and high brightness and contrast ratios. Thus, they are compatible with portable applications. High-information-content OLEDs using organic materials can be addressed using direct addressing, multiplex addressing or active matrix addressing. They are currently fabricated using electroluminescent low-molar-mass materials or aromatic conjugated electroluminescent polymers with a high glass transition temperature. Both of these classes of organic materials require a high T_g value in order to avoid crystallisation, which can degrade device performance severely. Indeed, life-time has been the major obstacle to commercialisation of this otherwise attractive technology. However, commercial OLEDs using low molecular weight materials and polymer are starting to appear on the flat-panel displays market in significant volumes.

Liquid Crystal Displays[11,12]

A variety of liquid crystal displays (LCDs) dominate the market for flat panel displays, especially for portable applications, see Table 1.1. The common features of these devices are low weight, thin planar construction, low operating voltages and power consumption, and acceptable contrast and viewing angles. LCDs can be operated in a reflection mode using ambient light, a transmission mode using a backlight and transflection combining both possibilities. LCDs invented more than 30 years ago were not the first successful portable flat panel display. Digital watches and calculators with segmented electrodes were first produced using LEDs incorporating electroluminescent inorganic semiconductors. However, LCDs soon displaced LEDs from these products and then enabled the fabrication, in the first instance, of digital watches and calculators, then notebooks and laptop computers, camcorder viewers, portable telephones, personal digital assistants, hand-held games, car navigation and orientation systems and many more applications. Surprisingly most of the original LCD prototypes were realised using nematic liquid crystals at temperatures over 100 °C. The design and synthesis of new organic compounds, which exhibit a nematic phase with a specific spectrum of properties, have been essential contributions in establishing and expanding the multibillion dollar LCD industry over the last 30 years as well as other multibillion dollar industries, *e.g.* mobile telephones and hand-held games, such as Gameboy, would have been much more difficult to bring to market without an LCD display. The optical performance of LCDs has been improved to such an extent that LCDs are starting to displace CRTs from applications where ergonomics and footprint and not power consumption, operating voltage and even cost are the deciding factors. However, the presence of absorbing polarisers in most types of LCDs and in all commercial LCDs manufactured at the moment, means that intrinsically, brighter emissive displays with Lambertian emission, such as OLEDs, are potential competitors in the FDP market for a spectrum of applications compatible with use in poor ambient light, if their performance can be improved to fulfil the specifications already met by LCDs. However, the memory effect of certain types of LCD, especially those with active matrix addressing, means that their power consumption can be very low. Emissive flat panel displays such as OLEDs continually draw power in the on-state and have to be driven continually from frame to frame like a CRT.

2 Conclusions

The flat-panel displays market is characterised by its diversity and increasing fragmentation. Therefore, many different display types dominate particular segments of the market where their particular combination of performance, system compatibility and cost represent the most appropriate choice. However, over the last decade the different types of LCD have become the dominant flat panel display for most applications. The market value of LCDs alone is estimated at approximately $13 000 000 in the year 2000. The market share by value of LCDs should overtake that of CRTs in the near future. However, the market for flat panel displays is expanding rapidly with the popularity of digital watches, calculators, notebook computers, personal digital organisers, palm-top computers, hand-held computer games and toys, mobile telephones, camcorders, digital cameras, *etc.* The rapid progress in mobile communications, especially those providing access to the internet and e-mail, will only serve to accelerate the growth in the market share and production volume of LCDs. The electro-optical performance of LCDs is continually being improved and the unit cost is steadily decreasing despite fluctuations due to economic dislocations and natural disasters in the Far East, *e.g.* in countries such as Japan, Korea, Hong Kong, Thailand and Malaysia, where most of the LCD manufacturing industry is based. Indeed the current low cost of many types of LCD is threatening the profitability of large parts of the LCD industry, which is particularly capital intensive. However, LCDs are starting to steadily displace the bulky CRT from crowded desktops, especially in countries such as Japan where space is at a premium. This trend can confidently be expected to continue and indeed accelerate in the near future.

The increased market share of LCDs is only slowing the volume growth of CRTs and competing flat-panel display technologies rather than reducing their overall volume or even displacing them from the displays' market place. Even the market for LCDs is remarkably diverse, although the falling cost of LCDs with active addressing (*e.g.* active matrix thin-film transistor (AM-TFT)-LCDs) is starting to reduce the market share of LCDs with multiplex addressing (super twisted nematic (STN)-LCDs) even from relatively low-cost applications. However, the number of display types of LCDs with active matrix addressing is increasing to meet the growing needs for light, large-area, high-information-content displays with video rate addressing, uniformly wide viewing angle and relatively high brightness.

The only flat-panel technology with the potential to pose a realistic challenge to LCDs in the medium term is OLED technology. The first factories for OLEDs using either small molecules or polymers have started production, if in relatively low volumes, see Table 1.1. Higher production volumes can be confidently expected as the market acceptability and awareness of the capability of OLEDs increases. A combination of the modulation of plane polarised light provided by an OLED back-light by an LCD to create a hybrid OLED-LCD may become a major commercial product in the near future. Oriented main-chain polymers or anisotropic polymer networks in the nematic liquid crystalline state have established themselves recently as the most attractive source of polarised electroluminescence. Moreover, liquid crystals in the smectic and columnar states have great potential as non-dispersive charge-carrier transport layers in multilayer OLEDs. Thus, the dominant flat-panel technology and its main competitor share many common elements, which could be combined to mutual benefit. Hybrid OLED-LCDs could expand the flat-panel displays market and the market share of OLEDs and LCDs simultaneously, *e.g.* for monochrome portable telephones with internet capability. The reasons for these prognoses will be illustrated in the following chapters with the emphasis on the dependence of these technologies on progress in research and development in organic materials chemistry.

3   References

1 C. Williams, *Electronic Materials for Displays Workshop*, London, UK, 1997.
2 J. Smith, *Proc. SID Euro Display '96*, 1996, 5.
3 J. Deschamps, *Proc. SID Euro Display '96*, 1996, 31.
4 K. Kasano and T. Nakamura, *Proc. SID Euro Display '81*, 1981, 156.
5 C. Williams, *Electronic Materials for Displays Workshop*, London, UK, 1997.
6 J. F. Peyre, *Proc. SID Euro Display '96*, 1996, 169.
7 L. J. Hornbeck, *SPIE Crit. Rev.*, 1989, **1150**, 86.
8 L. J. Hornbeck, *Int. Electron. Dev. Techn. Digest.*, 1993, 381.
9 R. H. Mauch, *Proc. SID Euro Display '96*, 1996, 601.
10 L. J. Rothberg and A. J. Lovinger, *J. Mater. Res.*, 1996, **11**, 3174.
11 T. Geelhaar, *Liq. Cryst.*, 1998, **24**, 91.
12 E. P. Raynes, *Proc. SID Euro Display '96*, 1996, 7.

CHAPTER 2

Liquid Crystals and Liquid Crystal Displays (LCDs)

1 Physical Properties of Nematic Liquid Crystals

The vast majority of LCDs incorporate a thin film of a mixture of organic compounds in the nematic phase, whose optical properties are modulated by an applied electric field.[1] These form the subject of this chapter and of Chapter 3.[2–18]

An LCD based on an electroclinic effect in achiral smectic A* phase, constituted of optically active rod-like molecules, was reported in the late 1970s. However, such LCDs have not been manufactured commercially despite intermittent interest and development, especially as spatial light modulators in applications for optical computing or telecommunications.[19] A very small number of commercial LCDs, which make use of the ferroelectric properties of liquid crystalline mixtures in the optically active chiral smectic C* phase to generate an electro-optical effect, are used as spatial light modulators, e.g. in optical computing or telecommunication applications or projection displays, e.g. for pilots' helmets.[20,21] Surface stabilised ferroelectric liquid crystal displays (SSF-LCDs) or other display types based on ferroelectric liquid crystals may eventually have the potential to overcome some of the limitations of nematic LCDs.[20,21] Analogous LCDs using antiferroelectric liquid crystals may also have the potential to break through into the FDP market.[22] However, LCDs using smectic liquid crystals are still manufactured on a very small scale, despite several decades of intense research and development. Therefore, LCDs using smectic liquid crystals will not be described further in this monograph. Laser-addressed LCDs for high-information-content applications, such as very detailed maps, which are based upon light scattering from side-chain liquid crystal polymers in the smectic A phase, have been manufactured on a very limited scale at a very high unit cost, especially for military and avionic applications. LCDs using ferroelectric side-chain liquid crystal polymers for lightweight displays, such as head-up displays for pilots, have also been developed. However, this type of LCDs utilising liquid crystal-line polymers have yet to find commercial acceptance in the consumer displays marketplace. Liquid crystalline polymers are described in depth elsewhere.[23,24]

Therefore, their applications in LCDs will not be discussed further in this monograph. They may find application as electroluminescent materials as a source of polarised light for LCDs, see Chapter 6. Columnar liquid crystals are also not used as the active switching element in LCDs, although they are used as passive optical compensation layers of negative birefringence for TN-LCDs and STN-LCDs in order to compensate for interference colours and to increase the effective viewing angle, see Section 7. They may also find application as organic charge-carrier transport layers in multilayer OLEDs, see Chapter 5.

In order to understand the basic principles of operation of the many different kinds of LCDs being developed and/or manufactured at the present time, it is necessary to briefly describe the liquid crystalline state and then define the physical properties of direct relevance to LCDs. First, the nematic, smectic and columnar liquid crystalline states will be described briefly. However, the rest of the monograph dealing with liquid crystals will concentrate on nematic liquid crystals and their physical properties, since the vast majority of LCDs manufactured operate using mixtures of thermotropic, non-amphiphilic rod-like organic compounds in the nematic state.

The unusual optical properties of liquid crystals had been remarked upon and described for several centuries before their uniqueness as a state of matter was recognised.[25–27] Their early reports described the strange melting behaviour and appearance of some naturally occurring materials, either as pure compounds or as gels in water, which have now been shown to be thermotropic or lyotropic liquid crystals. Thermotropic liquid crystalline phases are formed under the action of heat, see Figures 2.1 and 2.2, and the lyotropic liquid crystalline phases are formed by the action of a solvent, such as water, usually with an amphiphilic compound.[28,29] However, the nature of these materials, or indeed their exact

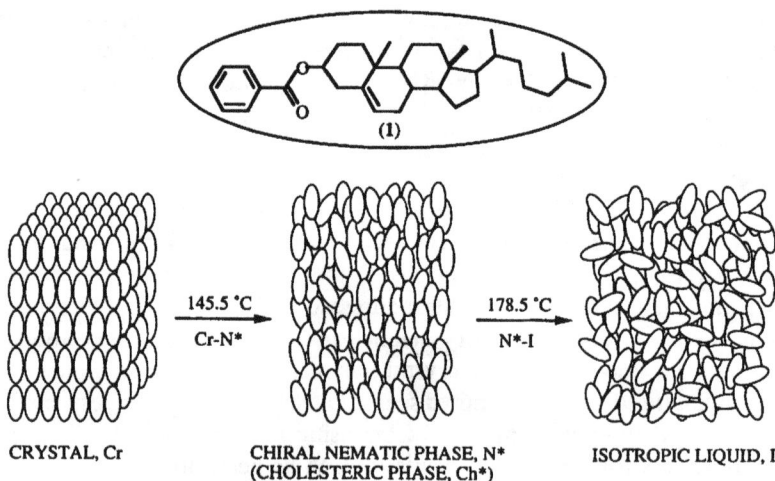

(1)

| 145.5 °C | 178.5 °C |
| Cr-N* | N*-I |

CRYSTAL, Cr   CHIRAL NEMATIC PHASE, N*   ISOTROPIC LIQUID, I
                           (CHOLESTERIC PHASE, Ch*)

Figure 2.1   *A schematic representation of the melting of crystalline cholesteryl benzoate (1) at 145.5 °C to form a chiral nematic (cholesteric) phase, which in turn forms the isotropic liquid on further heating to 178.5 °C.*[30,31]

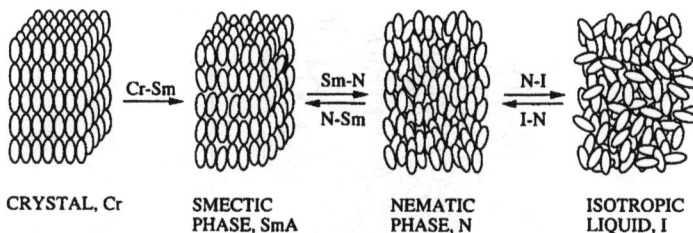

| CRYSTAL, Cr | SMECTIC PHASE, SmA | NEMATIC PHASE, N | ISOTROPIC LIQUID, I |

Figure 2.2 *Schematic representation of the structures of a solid, a smectic phase (SmA), the nematic phase and the isotropic liquid state formed by calamitic organic molecules with a large length-to-breadth ratio.*

chemical structure, was not known at the time. Therefore, the discovery of the liquid crystalline state is usually attributed to the botanist Reinitzer from the Institute for Plant Biology of the German University of Prague in 1888.[30] When determining the melting points in capillary tubes of cholesteryl benzoate (**1**), he noticed that it seemed to melt at 145.5°C to form a cloudy liquid with unusual reflection colours. This opaque liquid then appeared to melt again at 178.5°C to form a clear transparent liquid and then reform reversibly within a few degrees, see Figure 2.1. This liquid crystalline state is now known to be the cholesteric state (Ch), now referred to as the chiral nematic state (N*), *i.e.* the helical equivalent, formed by some optically active materials, of the usual nematic state (N), see Figure 2.3. The term 'fließende Kristalle' flowing crystals or liquid crystals was introduced[31] to describe this unusual state of matter with the fluidity of liquids and the optical properties of some crystals such as birefringence, see below. Liquid crystals are also often referred to as being mesomorphic, exhibiting mesophases, *i.e.* nematic, smectic or columnar mesophases or states 'in between' those of a crystal and a liquid.[32–37] A liquid crystal may exhibit only one of these states or several of them at different temperatures, see Figure 2.2. If the melting point of a particular compound is higher than the temperature of the transition from the isotropic liquid to the liquid crystalline phase then this phase is referred to as being monotropic. An enantiotropic phase is observed above the melting point (Cr–Sm, Cr–N, Cr–Col).

In the nematic state there exists a local parallel orientation of the molecular long axis of calamitic or rod-like molecules with a large length-to-breadth ratio, see Figures 2.1 and 2.2 and Table 2.1.[38–43] This orientation may extend over several micrometres. This parallel orientation of the molecular long axes is absent in the liquid state, which may be considered as completely isotropic even for rod-like molecules, see Figures 2.1 and 2.2. This is described by the continuum theory, which postulates that the director of the nematic state changes continually and gradually throughout the bulk of the nematic material.[44–50] The average orientation of the molecular long axes at any one point is defined as the director, n. There is an inversion symmetry axis at any one point along the director and, therefore, n is identical to $-n$. The order parameter, S, is a macroscopic scalar quantity, which represents the average orientation of the molecules in the liquid crystal relative to the director:[51]

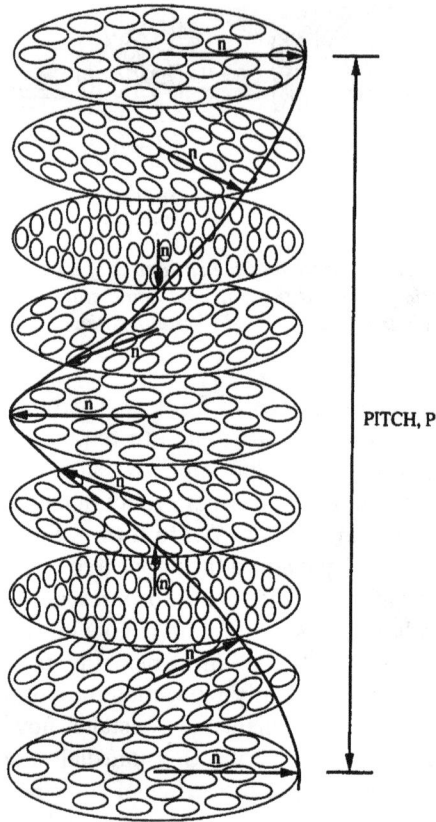

Figure 2.3 *Schematic representation of the periodical helical structures of the chiral nematic (cholesteric) phase. The pitch of the helix corresponds to the rotation of the director through 360°. There is no layered structure in a chiral nematic, N*, phase.*

$$S = \left\langle \frac{1}{2}\left(3\cos^2\theta - 1\right)\right\rangle \tag{1}$$

where θ is the angle between the long molecular axis of an individual molecule and the director. In a completely disordered liquid made up of rod-like molecules

$$\cos^2\theta = \frac{1}{3} \tag{2}$$

and S is 0. In an ideal macroscopically ordered nematic liquid crystalline state

$$\cos^2\theta = 1 \tag{3}$$

and $S = 1$. For a typical compound in the nematic state at a temperature

Table 2.1 *Transition temperatures (°C) for the nematic liquid crystals 2–16*

	Molecular structure	Cr		N		I	Ref
2	C$_4$H$_9$———OCH$_3$	●	69	–		●	38
3	C$_4$H$_9$——O——OCH$_3$	●	<25	–		●	38
4	C$_4$H$_9$——≡——OCH$_3$	●	49	(●	37)	●	31,38
5	C$_4$H$_9$——N(H)—C(O)——OCH$_3$	●	145	–		●	38
6	C$_4$H$_9$——C(O)—O——OCH$_3$	●	61	(●	25)	●	39
7	C$_4$H$_9$——O—C(O)——OCH$_3$	●	40	(●	25)	●	39
8	C$_4$H$_9$——C(H)=C(H)——OCH$_3$	●	118	●	121	●	38,39
9	C$_4$H$_9$——N=N——OCH$_3$	●	32	●	47	●	39,41
10	C$_4$H$_9$——C(H)=N$^+$(O$^-$)——OCH$_3$	●	108	(●	70)	●	38
11	C$_4$H$_9$——N$^+$(O$^-$)=C(H)——OCH$_3$	●	113	(●	53)	●	38,42
12	C$_4$H$_9$——N=N$^+$(O$^-$)——OCH$_3$	●	42	●	77	●	38,42
13	C$_4$H$_9$——N$^+$(O$^-$)=N——OCH$_3$	●	41	●	74	●	38

(continued)

Table 2.1 *continued*

	Molecular structure	Cr		N		I	Ref
14	C₄H₉―⟨⟩―C(=N―⟨⟩―OCH₃)H	●	46	●	49	●	42
15	C₄H₉―⟨⟩―N=C(H)―⟨⟩―OCH₃	●	20	●	47	●	43
16	C₄H₉―⟨⟩―OCO―⟨⟩―OCH₃	●	28	–		●	38

Parentheses represents a monotropic transition temperature.

relatively far away from the clearing point, T_{N-I}, *i.e.* the temperature at which the compound ceases to exhibit the nematic state and forms the liquid state, usually S lies between 0.5 and 0.7. The Maier–Saupe theory postulates that the order parameter, S, at a temperature T depends on the reduced temperature T_{red}, (see Figure 2.4):

$$T_{red} = \frac{T}{T_{N-I}}$$

The direction of the director is random in space and time. However, the director

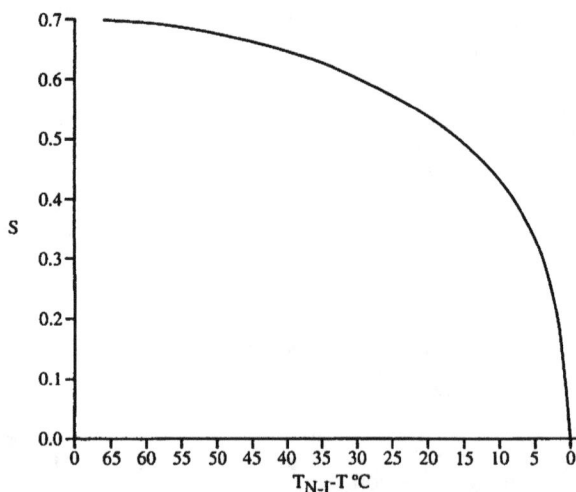

Figure 2.4 *The dependence of the order parameter, S, on the reduced temperature, T_{N-I}-T, where T_{N-I} is the clearing point (i.e. the temperature at which the transition from the nematic phase into the isotropic liquid takes place) and the measurement temperature, T.[51]*

of a bulk sample of a material in the nematic phase can be macroscopically and almost uniformly oriented in one direction by a relatively weak external force such as an electric or magnetic field, see Section 2, or an aligning surface, see Section 6. The bulk realignment of the director, and thus the optic axis, by surface forces and electric fields represents the underlying electro-optical effects of liquid crystal displays using nematic liquid crystals, see Sections 3–7 and Chapter 3. If a compound exhibits a nematic phase and one or more smectic phases, the nematic phase is nearly always exhibited at a temperature above that of the smectic phases due to the higher degree of order in smectic mesophases, see Figure 2.2.

In the twelve smectic phases identified to date the molecules are arranged in layers, see Figure 2.2.[33–37] The physical properties of these smectic phases depend on the ability of the layers to slide over one another and to bend, the average orientation of the director within the layers and the degree of order within and between the layers. The director is orthogonal to the layer plane in several smectic phases (SmA, SmB [hexatic] and SmE) and is tilted with respect to the layer normal in some other smectic phases (SmC, SmF and SmI). The more ordered phases (SmB, SmF and SmI) with hexagonal repeat units within the layers are exhibited at lower temperatures. Some smectic phases with long-range correlations between molecules in individual layers are more properly referred to as plastic crystals (Crystals B, E, J, G, K and H). There are also a large and increasing number of antiferroelectric and ferrielectric smectic phases. Optically active compounds can also exhibit the chiral versions of the smectic phases described above, denoted by an asterix, *e.g.* SmC*. The layer structure of smectic phases results in a large bulk flow viscosity. This restricts their use in those electro-optical display device configurations, which now use the nematic phase, due to the long response times. However, a number of different types of LCD using compounds in various smectic phases have been developed. These are referred to above and do not form part of this monograph due to space considerations and their very low share of the commercial LCD market.

Columnar liquid crystalline phases are formed generally by disc-shaped molecules or self-assembled aggregates, see Figure 2.5. These are organised in a supramolecular structure of nearly parallel columns with different degrees of order in a two-dimensional lattice, see Section 4 of Chapter 5.[36,37] There is no regular columnar structure in the nematic columnar phase formed at higher temperatures. In columnar mesophases the molecular cores are organised above each other in columns separated by the terminal chains on the outside of the columns. Therefore, the intercolumnar distance (≈ 15–40 Å) is much greater than the intracolumnar distance (< 4.5 Å) depending on the length and conformation of the aliphatic chains. There are columnar phases with hexagonal and rectangular lattices (*i.e.* columnar hexagonal, col_h; rectangular, col_t; oblique, col_{ob}) which are either ordered (*e.g.* col_{ho}) or disordered (*e.g.* col_{hd}). The nature of columnar phases, in which the aromatic cores of the molecules are arranged above each other in columns, allows efficient charge-carrier migration due to the overlap of the π-electron orbitals in the conjugated aromatic core of

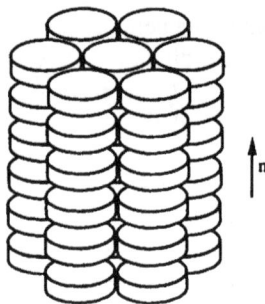

Figure 2.5 *Schematic representation of the D_{ho} columnar phase with an ordered hexagonal arrangement of the columns of disc-shaped molecules with a regular period of the discs within the columns. The director is parallel to the columns and normal to the plane of the discs.*

neighbouring molecules. This can be made use of in organic light-emitting diodes (OLEDs), see Chapter 5, Section 4 and photovoltaic effects.

Many new synthetic liquid crystals, which were found to exhibit smectic and/ or nematic mesophases, were prepared from the beginning of the 20th century, especially at the Martin Luther University in Halle in Germany, and in the United Kingdom at the University of Hull from the early 1950s to the present day. These systematic studies served to establish the relationships between molecular structure and the type of liquid crystalline state (mesophase) exhibited, *i.e.* nematic, smectic or columnar, and at what temperatures.[32–37] Some examples of the molecular structures of typical nematic liquid crystals (2–16) synthesised up to the late 1960s are shown in Table 2.1.[38–43] A typical liquid crystal of this time possessed a linear structure with a central core containing several collinear rings, a linear unsaturated linkage and two terminal chains. The combination of one short alkyl or alkoxy chain on one ring and a polar substituent on a second aromatic ring was found to promote nematic phase formation. All of the compounds reported in Table 2.1 are aromatic with two phenyl rings joined by an unsaturated central linkage. This seems to suggest that the formation of the liquid crystalline state is promoted by a high degree of conjugation and a large anisotropy of molecular shape and polarisability. These findings are consistent with the basic theories describing the factors responsible for the formation of the liquid crystalline state and especially the nematic state.[44–55]

Calamitic compounds which exhibit a smectic and/or nematic phase usually consist of a relatively rigid central core containing co-linear six-membered rings, either aromatic rings, such as 1,4-disubstituted-phenylene, 2,5-disubstituted-pyridine, 2,5-disubstituted-pyrimidine, 3,6-disubstituted-pyridazine, and alicyclic rings, such as *trans*-1,4-disubstituted-cyclohexane, 1,4-disubstituted-bicyclo[2.2.2]octane, 2,5-disubstituted-dioxane. Heteroaromatic rings tend to lead to the formation of smectic phases rather than the nematic phase unless combined with a polar terminal function, such as a cyano group. The dependence of the liquid crystalline transition temperatures on the nature of

Table 2.2 *Transition temperatures (°C) of the nitriles 17–23*

	Molecular structure	Cr		N		I
17	C_5H_{11}—⬡—⬡—CN	●	22.5	●	35	●
18	C_5H_{11}—⬡—⬡—CN	●	31	●	55	●
19	C_5H_{11}—⬡—⬡—CN	●	62	●	100	●
20	C_5H_{11}—⬡—⬡—CN	●	<25	●	−25	●
21	C_5H_{11}—⬡—⬡—CN	●	113	(●	50)	●
22	C_5H_{11}—⬡—⬡—CN	●	62	●	85	●
23	C_5H_{11}—⬡—⬡—CN	●	104	●	129	●

Parentheses represent a monotropic transition temperature.

the six-membered rings is shown by the data for the two-ring compounds (17–23) collated in Table 2.2. The sequential replacement of the phenyl rings in the fully aromatic compound (17) by the saturated alicyclic cyclohexane and bicyclo[2.2.2]octane rings to yield the compounds 18–23 leads to a systematic increase in the nematic clearing point. It is evident that the presence of a phenyl ring and a non-conjugated cyano group in compound 19 gives rise to a low clearing point. This illustrates the fact that it is very difficult to extrapolate from one compound to another. This is dealt with in much more detail in Chapter 3.

Compounds containing 2,5-disubstituted five-membered rings, such as furan, thiophene, 1,3,4-oxadiazole and 1,3,4-thiadiazole, usually exhibit lower nematic clearing points than those of similar compounds containing six-membered rings. This is due to the non co-linear nature of the bonds of these five-membered rings. Materials containing only two rings in the core are not usually liquid crystalline if one of them is a five-membered ring. Condensed 2,6-disubstituted rings, such as naphthalene, quinoline, quinoxoline, tetralene, chromane, dioxynaphthalene and *trans,trans*-decalin, do not give rise to mesophase formation in the absence of another ring in the molecular core. This is due to the small length-to-breadth ratio of such fused rings. Compounds containing alicyclic or aromatic rings with more than six units, such as cycloheptane, cycloheptatrienone, diazazulene and tropone, are generally not mesomorphic, unless two other six-membered rings are also present. Two six-membered rings are usually required in a molecule for the formation of a mesophase. Carboxylic acids with only one ring, such as 4-methoxycinnamic

Table 2.3 *Typical values[1] and symbols for some physical properties of 4-cyano-4'-pentylbiphenyl (17)[36,37] and 4-(trans-4-pentylcyclohexyl)benzonitrile (18)[58,59]*

Property and symbol	Value/Units K15 (17)	Value/Units PCH5 (18)
Melting point (Cr-N)	22.5 °C	30 °C
Clearing point (N-I)	35 °C	55 °C
Melting point enthalpy	4.1 kcal mol^{-1}	21.35 kJ mol^{-1}
Clearing point enthalpy		0.96 kJ mol^{-1}
Density (ρ)	1.019 g cm^{-3} mol$^{-1\,a}$	0.9706 g cm$^{-3\,b}$
Bulk viscosity (η)		22.5 mm s$^{-1\,b}$
Rotational viscosity (γ_1)		0.1507 Pa s^{b}
Refractive index of the extraordinary ray (n_e)	1.702^{a}	1.6173^{b,c}
Refractive index of the ordinary ray (n_o)	1.539^{a}	1.4924^{b,c}
Birefringence (Δn)	0.163^{a}	0.1249^{b}
Parallel dielectric permitivity ($\varepsilon_\parallel$)	17.1^{a}	17.5^{b,d}
Orthogonal dielectric permitivity ($\varepsilon_\perp$)	7.2^{a}	4.8^{b,d}
Dielectric anisotropy ($\Delta\varepsilon$)	9.9^{a}	12.7
Splay elastic constant (k_{11})		9.6 10^{-12} N^{b}
Twist elastic constant (k_{22})		6.5 10^{-12} N^{b}
Bend elastic constant (k_{33})		19.4 10^{-12} N^{b}
Bend/splay elastic constant ratio (k_{33}/k_{11})	1.46	2.03
Bend/twist elastic constant ratio (k_{33}/k_{22})		3.0
Diamagnetic anisotropy (χ)	1.51 10^{-9} m^3 kg^{-3}	0.46 10^{-9} m^3 kg^{-3}

[a] Measured at 29 °C; [b] measured at 20 °C; [c] measured at 546 nm; [d] measured at 1 kHz.

acid, can exhibit a liquid crystalline phase. However, this is due to the formation of molecular dimers due to hydrogen bonding between the carboxylic acid groups. This results in a dynamic equilibrium of a mixture of one-ring and quasi three-ring dimers, which possesses a nematic phase.

The rings in the molecular core may be linked directly or by a conjugated linkage with two units, such as carboxy (-CO$_2$-), ethenyl (-CH=CH-), ethynyl (-C≡C-), azo (-N=N-), azoxy (-N(O)=N-), azomethine (-CH=N-), or an aliphatic linkage with two units, such as ethyl (-CH$_2$CH$_2$-) or methylenoxy (-CH$_2$O-), see Table 2.1. Linkages with four units, such as butyl, (*E*)-butenyl, propyloxy and (*E*)-propenyloxy have also been used but much more rarely. Linkages with an odd number of units, such as methyl or propyl, lead to a non-linear structure. Consequently, the clearing point of these compounds is usually very low.

At least one terminal chain, such as an alkyl (R) or alkoxy chain (RO) is attached to the molecular core. The second group may be a similar group or a small substituent such as a halogen atom (F, Cl, Br, I) or a cyano (CN), isothiocyanato (NCS), dimethylamino (N(CH$_3$)$_2$) or nitro (NO$_2$) group or a fluorinated group such as trifluoromethyl (CF$_3$), trifluoromethoxy (CF$_3$O), difluoromethoxy (CF$_2$HO), *etc.* Compounds with two short terminal alkyl or alkoxy chains tend to exhibit a nematic phase. Mesomorphic compounds with short

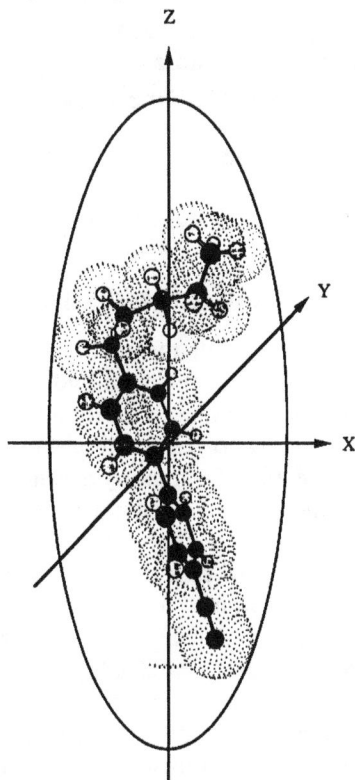

Figure 2.6 *The molecular structure of a typical calamitic, thermotropic nematic liquid crystal. The diameter of the molecule measured in the x–y plane is symmetrical about the z-axis due to very rapid rotation about the molecular long axis.*

terminal chains usually exhibit a nematic phase. However, most liquid crystalline materials with long terminal chains, *e.g.* 7–12 methylene units, often exhibit a smectic phase. Mesomorphic compounds with longer chains generally only possess smectic phases. A general order of increasing tendency of these terminal groups to induce nematic phase formation has been established empirically, as shown below, although these do vary for different molecular structures:

$$RO > R > CN > OCH_3 > NO_2 > Cl > Br > N(CH_3)_2 > CH_3 > I > CF_3 > H$$

Compounds with an odd number of methylene (CH_2) or oxygen atoms in the chain usually exhibit a higher clearing point than that of homologues with an even number of units in the chain, *i.e.* an odd–even effect is often observed for the nematic clearing point of a homologous series. The shapes of the plots of nematic clearing point against number of units in the terminal shape are regular, but may exhibit a variety of patterns. The same groups in a lateral, *i.e.* non-terminal, position lead to a lower melting point and clearing point due to a broadening of the molecular rotation volume, *i.e.* the length-to-breadth ratio is reduced. Smectic phases are also suppressed. This can lead to a broader nematic phase at lower temperatures than those of corresponding materials without a lateral substituent. Chain-branching can give rise to lower transition temperatures for similar reasons.

The correlation between molecular structure, liquid crystal transition temperatures and physical properties of the nematic phase of these materials of relevance to individual types LCDs is dealt with extensively in Chapter 3.

2 Physical Properties of Liquid Crystals[1,15,25,56–58]

The rod-like shape of liquid crystals means that their physical properties are anisotropic, *i.e.* they are of a different magnitude when measured parallel or perpendicular to the director,[1] see Figure 2.6 for a molecular model of a typical nematic liquid crystal, *i.e.* 4-cyano-4'-pentylbiphenyl. Free rotation about the molecular long axis gives an axis of symmetry parallel to the director so that the values of the physical properties measured normal to the director, *i.e.* along the y- and z-axes are identical. However, they differ from those measured parallel to the director, *i.e.* along the x-axis. It is the anisotropic nature of the physical properties of liquid crystals combined with the ability of magnetic and electric fields to quickly influence the spatial orientation of the director, and thus the optic axis, which enables electro-optical display devices to be constructed. The fast reorientation of the director under the influence of a moderate electric field is a result of their fluid nature and low viscosity. The physical properties are temperature and pressure dependent as well as depending on the type of liquid crystal state, *e.g.* nematic, smectic, columnar and the degree of order in that state. The most important anisotropic properties of the nematic state of relevance to LCDs are described below (see Table 2.3).[1,15,25,56–58]

Optical Anisotropy (Birefringence)

The nematic phase of macroscopically aligned calamitic liquid crystals is uniaxial due to the anisotropy of shape and polarisation. They are, therefore, also optically anisotropic, *i.e.* birefringent, in that they exhibit different refractive indices for light travelling parallel and perpendicular to the director (optic axis). The physical property of birefringence is manifested by certain crystalline solids, *i.e.* transparent crystals with a non-centrosymmetrical lattice structure, such as calcite. The complete freedom of rotation in liquids averages out to zero any anisotropic molecular properties and renders the bulk liquid optically isotropic. Incident plane polarised light entering a birefringent medium, such as a non-centrosymmetrical transparent crystal or a thin transparent film of a macroscopically aligned nematic phase, is split into two mutually perpendicular components called the ordinary (o) and extraordinary (e) rays. The electric field of the o-ray is perpendicular to the optic axis. Therefore, the refractive index, n_o, of the medium for the o-ray is a constant independent of propagation direction of the ray. The electric field of the e-ray lies in a plane containing the optic axis of the medium. Therefore, the effective

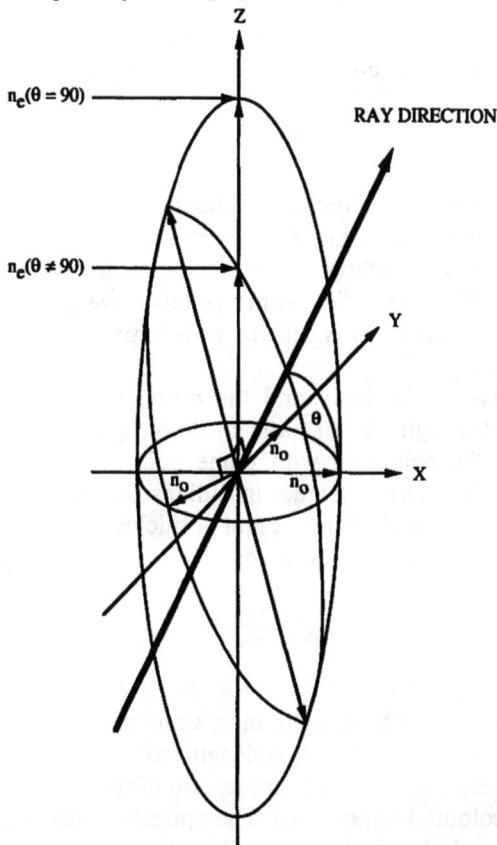

Figure 2.7 *The refractive index ellipsoid of a uniaxial liquid crystal phase with the optical axis parallel to the x-axis. The refractive index, n_o, of the ordinary ray is independent of the direction of propagation. The refractive index, n_e, of the extraordinary ray is larger than n_o if the liquid crystalline phase is of positive birefringence.[1]*

refractive index, $n_e(\theta)$, of the medium for the e-ray depends on the angle, θ, the ray makes with respect to the optic axis (see Figure 2.7):

$$n_e(\theta)^2 = \left(\frac{\cos^2\theta}{n_o^2} + \frac{\sin^2\theta}{n_e^2}\right)^{-1} \tag{5}$$

A consequence of these different refractive indices for the ordinary ray and the extraordinary ray is that the effective birefringence of the medium, $\Delta n(\theta)$, also depends on the propagation direction:

$$\Delta n(\text{th}) = n_e(\theta) - n_o \tag{6}$$

whereby the maximum value of the birefringence, Δn, is found for $\theta = 90°$, *i.e.* when the electric field of the e-ray is parallel to the optic axis. In this case

$$\Delta n = n_e - n_o \tag{7}$$

where $n_e = n_e(\theta = 90°)$. Therefore, Δn is the difference between the refractive indices for the o-ray and the e-ray of a macroscopically aligned nematic phase propagating parallel and orthogonal, respectively, to the optical axis of the nematic medium. Most nematic liquid crystals have positive birefringence, $\Delta n > 0$, meaning that the e-ray is delayed with respect to the o-ray on passage through the material.[1,59,60]

Interference between the e-ray and the o-ray, which have travelled with different velocity through the nematic medium, gives rise to the coloured appearance of LCDs operating with plane polarised light. For a wave at normal incidence, the phase difference in radians between the o-ray and the e-ray caused by traversing a birefringent film of thickness d and birefringence Δn, is referred to as the optical retardation, δ:

$$\delta = \frac{2\pi \Delta n d}{\lambda_v} \tag{8}$$

where λ_v is the wavelength of light in a vacuum. The amount of optical retardation is dependent on the wavelength so that interference occurs at different frequencies resulting in the suppression of some of the visible spectrum and, therefore, a coloured appearance. The optical retardation depends on the path length, d, through the display. Therefore, the interference colours observed are also dependent on the viewing angle, since d will vary with θ. The birefringence and the optical retardation are also dependent on the wavelength of light and temperature, since the magnitude of the refractive indices also vary with these parameters. The material is no longer birefringent in the isotropic liquid above the nematic clearing point ($n_e = n_o$). Consequently, an isotropic refractive index, n_i, is observed, see Figure 2.8.

Elastic Constants[1,46]

The spatial and temporal response of a nematic phase to a distorting force, such as an electric (or magnetic) field is determined in part by three elastic constants, k_{11}, k_{22} and k_{33}, associated with splay, twist and bend deformations, respectively, see Figure 2.9. The elastic constants describe the restoring forces on a molecule within the nematic phase on removal of some external force which had distorted the nematic medium from its equilibrium, *i.e.* lowest energy conformation. The configuration of the nematic director within an LCD in the absence of an applied field is determined by the interaction of very thin layers of molecules with an orientation layer coating the surface of the substrates above the electrodes. The direction imposed on the director at the surface is then

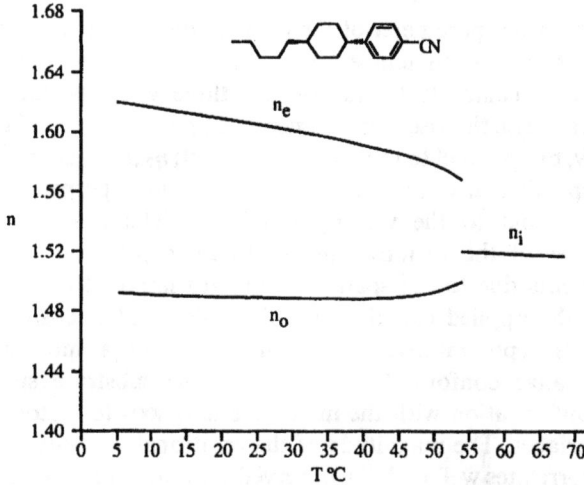

Figure 2.8 *The dependence of the refractive indices, n_o and n_e, of the ordinary and extraordinary rays, respectively, on the temperature, T, for a typical nematic liquid crystal. Above the clearing point, T_c, there is no birefringence and only one refractive index, n_i, is observed.*[58]

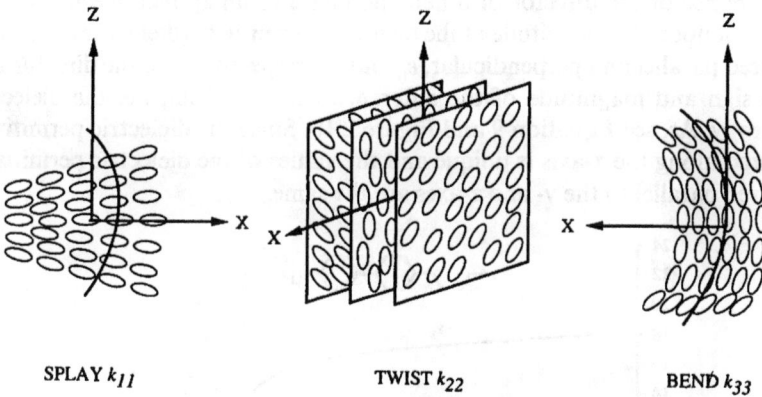

SPLAY k_{11} TWIST k_{22} BEND k_{33}

Figure 2.9 *Schematic representation of the elastic constants for splay, twist and bend, k_{11}, k_{22} and k_{33}, respectively, of a nematic phase.*[1,58]

transmitted to the bulk of the nematic phase by elastic forces, see Section 6 and Chapter 3.

Viscosity[1,61]

The flow viscosity of a nematic phase also determines the spatial and temporal response of the director to an applied field. The bulk viscosity of a nematic phase depends on the direction of flow of each molecule with respect to the director, averaged out over the whole of the sample. Therefore, bulk viscosity is dependent on the order parameter of the nematic phase and, consequently, its magnitude is higher at lower temperatures. However, due to the anisotropic molecular shape of calamitic liquid crystals, three viscosity coefficients are required to characterise the viscosity of a nematic phase: η_1 perpendicular to the direction of flow, but parallel to the velocity gradient; η_2 parallel to the direction of flow, but perpendicular to the velocity gradient; and η_3 perpendicular to the direction of flow and to the velocity gradient.[61] The response times and operating voltages of the various types of LCDs depend on the individual viscosity coefficients due to the spatial asymmetry imposed by the boundary conditions and the applied electric field. The rotational viscosity, γ_1, of the nematic phase is representative of the movement of a molecule from a homogeneous planar conformation parallel to the substrate surfaces to a homeotropic conformation with the molecular long axis (director) normal to the substrate surfaces. The magnitude of the rotational viscosity of a nematic medium often correlates well with the observed response times in LCDs, *i.e.* the higher the viscosity the longer the response time under a given set of driving conditions.

Dielectric Anisotropy

The response of the director of a nematic phase to an applied electric field is dependent upon the magnitude of the dielectric permitivity (dielectric constants) measured parallel and perpendicular, $\varepsilon_{||}$ and $\varepsilon_{\perp}$, respectively, to the director and to the sign and magnitude of the difference between them, *i.e.* the dielectric anisotropy, $\Delta\varepsilon$, see Equation 9 and Figure 2.10. Since the dielectric permitivity measured along the x-axis is unique and the values of the dielectric permitivity measured parallel to the y- and z-axes are the same,

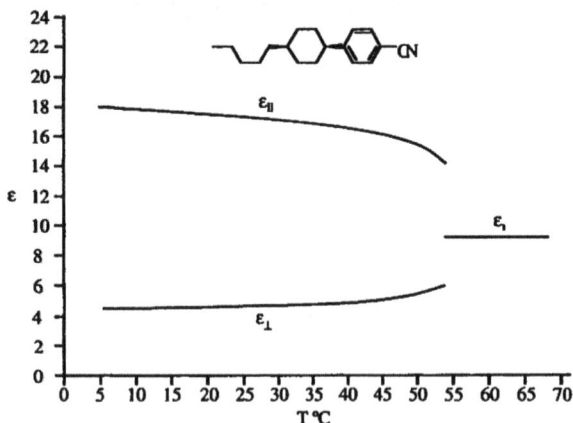

Figure 2.10. *The dependence of the dielectric constants, $\varepsilon_{\perp}$ and $\varepsilon_{||}$, measured perpendicular and parallel, respectively, to the nematic director, on the temperature, T, for a typical nematic liquid crystal. Above the clearing point, T_c, the dielectric anisotropy, $\Delta\varepsilon = \varepsilon_{||}-\varepsilon_{\perp}$, disappears and only one dielectric constant, ε_i, the permitivity of the isotropic liquid, is observed.*[1,58]

$$\Delta\varepsilon = \varepsilon_{\parallel} - \varepsilon\perp \qquad (9)$$

The dielectric permitivity, ε, of a dielectric, *i.e.* insulating, material is the ratio of the capacitance, C_{mat}, of a parallel plate capacitor containing that material to the capacitance, C_{vac}, of the same capacitor containing a vacuum:

$$\frac{C_{mat}}{C_{vac}} = \varepsilon \qquad (10)$$

The dielectric constants of an aligned nematic phase are dependent upon both the temperature and the frequency of the applied field at temperatures below the clearing point. The dielectric permitivity, ε_i, measured parallel to all three axes above the clearing point in the isotropic liquid is the same. Therefore, the dielectric anisotropy of the same compound in the liquid state is zero, see Figure 2.10. The sign and magnitude of the dielectric constants and, therefore, the dielectric anisotropy are dependent upon the anisotropy of the induced molecular polarisability, $\Delta\alpha$, as well as the anisotropy and direction of the resultant permanent molecular polarisation determined by permanent dipole moments.

3   Liquid Crystal Displays[2–18]

Use has been made of three distinct physical phenomena associated with the nematic state in order to modulate the passage of light through a thin film of a material in the nematic state by an electric field.[2–18] These electro-optical effects have only been used since the late 1960s to create a wide variety of different types of LCDs, see Chapter 3, although the interaction between the nematic state and magnetic and electric fields had been investigated since the beginning of the 19th century.[2–5]

The first LCD was reported by Heilmeier, Zanoni and Barton at the RCA Laboratories in Princeton, New Jersey in 1968[62,63] and made use of an electrohydrodynamic effect in a nematic medium under the action of an applied electric field, although changes in the orientation of the director due the stirring effect of an electric field on the nematic state had already been observed in the 1930s.[3] The formation of visible domains, known as Williams domains, at low voltages and light scattering at higher threshold voltages had also been discovered at the RCA Corporation several years earlier.[5,64] The optical effect is due to electrohydrodynamic instabilities generated by the movement under the action of an electric field of ions in the nematic phase.[65,66] The movement of the ions under the action of an electric field results in the circular movement of adjacent liquid crystal domains in different directions to produce visible stripes referred to above as Williams domains.[5,64] The resultant shear is balanced out by the elastic and dielectric torques under steady state conditions at low applied voltage. However, at higher fields these visible black-and-white stripes are replaced by a bright white appearance as incident visible light is scattered homogeneously. This effect is referred to as dynamic scattering and formed the underlying electro-optical effect of the first LCD to be manufactured on a commercial scale as a flat panel display for portable instruments. LCDs based on dynamic scattering are no longer manufactured due to various reasons, see Chapter 3.

The second physical effect made use of to fabricate an LCD was a field effect, *i.e.* the reorientation of the nematic director and, therefore, the optic axis, under the action of an electric field. Organic compounds of high resistivity are insulating dielectrics in the nematic state. Therefore, the application of an electric field results in a reorientation of the nematic director either parallel or perpendicular to the direction of the applied field due to the dielectric coupling of the field with the induced and permanent dipole moments of the molecules.[1,56,57] The molecules align themselves with the molecular axis of greatest resultant polarisation parallel with the field. The dielectric realignment of the director of a nematic phase by an electric field can be described essentially by the same equations as those used to define the effect of magnetic field on a nematic liquid crystal with appropriate modifications due to the different nature of the applied field. This corresponds to the Kerr effect observed in liquids, but is of a much higher magnitude in the nematic state. This is the basic electro-optical effect used in all commercial LCDs using nematic liquid crystals being manufactured at the moment.[2-18]

The third electro-optical effect using calamitic nematic liquid crystals makes use of a flexoelectric effect manifested by a curved asymmetrical nematic medium. This corresponds to piezoelectricity in crystals. The existence of flexoelectricity in a nematic phase under certain boundary conditions was predicted in the late 1960s[67] and then confirmed experimentally several years later.[68] However, LCDs using this effect, such as bistable nematic displays[69-71] are only in the development stage and as such they will not be discussed in this monograph.

If polymerisation of a reactive monomer takes place in a mixture of the monomer and a non-polymerisable liquid crystalline medium, then very many small droplets of the unreactive nematic mixture are formed by phase separation from the polymer as the polymerisation progresses. These droplets can then be addressed by an external electric field. Such mixtures can be polymerised to form thin sheets. If the refractive indices of the polymer and the nematic phase are well matched, then the device appears transparent in the on-state and opaque in the off-state due to light scattering, if the size of the droplets is larger than the wavelength of visible light, *e.g.* 1 μm. This is the physical basis of polymer dispersed (PD)-LCDs.[72-74] These LCDs are commercially available and are used as coatings in privacy screens and many other applications. Gel-stabilised LCDs are strongly related to PD-LCDs. Anisotropic gels are also formed from a reactive monomer and a liquid crystalline mixture. However, the monomer is usually also a liquid crystal. This forms a liquid crystalline polymer network on polymerisation. The liquid crystalline mixture occupies the space within the network.[75-81] These liquid crystalline gels are much more stable to external shocks since coupling between the immobile network and the low-molar-mass liquid crystal molecules stabilises their orientation. In principle all

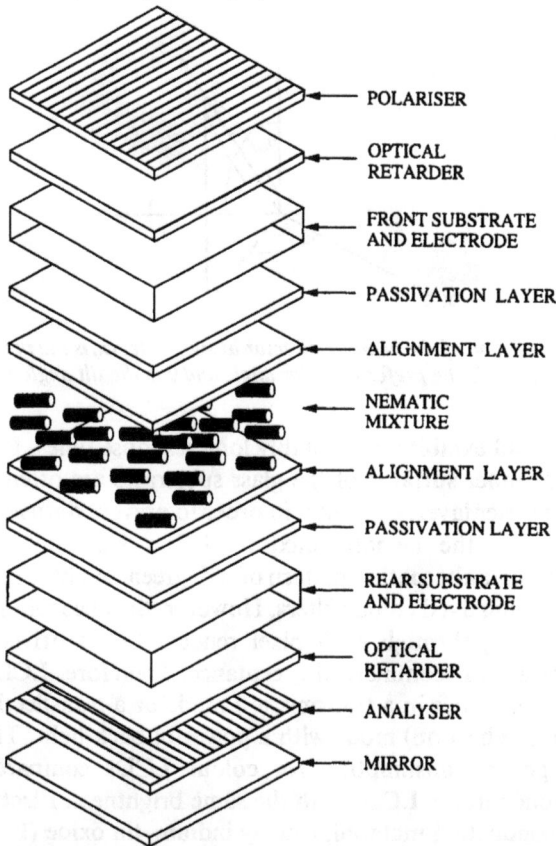

POLARISER

OPTICAL RETARDER

FRONT SUBSTRATE AND ELECTRODE

PASSIVATION LAYER

ALIGNMENT LAYER

NEMATIC MIXTURE

ALIGNMENT LAYER

PASSIVATION LAYER

REAR SUBSTRATE AND ELECTRODE

OPTICAL RETARDER

ANALYSER

MIRROR

Figure 2.11 *A schematic representation of the elements of a generalised liquid crystal display (LCD).*

the commercial LCD types described in Chapter 3 can be produced as a gel-stabilised LCDs or polymer-dispersed LCDs. However, the electro-optical effect and the fundamental modes of operation are very similar.[75-81] Therefore, they will not be discussed further in this monograph.

4   Cell Construction of Liquid Crystal Displays

The basic modes of construction of the various commercial types LCD described in detail in Chapter 3 share many common elements. They consist of a very thin layer of a nematic liquid crystal mixture enclosed between two transparent parallel glass substrates held apart by solid spacers and glued together around the edges, see Figure 2.11 for a schematic representation of the optical elements, some or all of which can be combined to construct an LCD. The cell gap ($d \approx 2$–10 μm) should be as uniform as possible in order to minimise variations in the optics of the display. Thick cells are turbid, due to

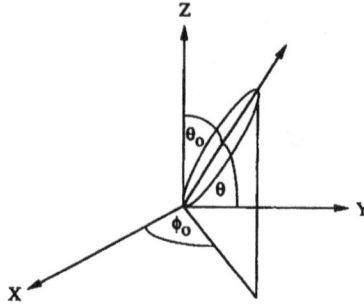

Figure 2.12 *Orientation of the nematic director at substrate. ϕ_0 is the preferred azimuthal angle, θ_0, is the preferred polar angle and θ is the tilt angle commonly used.*

light scattering, and exhibit unacceptably long response times. Before the LCD is assembled the inner surfaces of the glass substrates are often coated with a passivation or barrier layer, *e.g.* silica, in order to prevent diffusion of ions from the substrates into the nematic mixture. LCDs with full colour usually incorporate a regular alternating pattern of red, green and blue picture elements (pixels) formed using dyed colour filters. However, the absorption of two thirds of the light passing through each pixel renders LCDs driven in reflection insufficiently bright for commercial acceptance. Therefore, LCDs with colour filters are operated either in a transmission mode or a transflection (combined transmission and reflection) mode with a powerful back-light. This results in a much higher power consumption for colour LCDs compared to that of comparable monochrome LCDs with the same brightness. Electrodes made of a transparent conducting material, usually indium–tin oxide (ITO), are deposited on top of the these layers. Another thin barrier layer intended to prevent diffusion of ions into the nematic mixture may be deposited on top of the electrodes. This is followed by an alignment layer, which will be in direct contact with the nematic mixture in order to induce a homogeneous orientation of the director in the azimuthal and zenithal plane of the device, see Figure 2.12. The two glass substrates are then assembled and glued together leaving a hole so that the evacuated cell can be filled with a nematic liquid crystal under positive pressure, cleaned and then sealed, *e.g.* with epoxy resin or gold. The glass substrates may be offset in order to allow the drive electronics to be connected to LCDs with direct addressing or multiplexed addressing. The contacts are usually sheets of plastic coated alternating strips of a conducting polymer and an insulating polymer. The diameter of the strips correspond to the width of the row or column electrodes on LCDs with multiplex addressing.

Direct, multiplex and active matrix addressing are the three electronic drive methods used to generate the appropriate voltage at a particular pixel of an LCD, see Figures 2.11–2.14. The size, shape and pattern of electrodes on LCD substrates are fashioned to be compatible with the chosen method of addressing. In directly addressed LCDs the desired pattern of pixel electrodes is created by etching on one surface. A non-patterned back electrode on the second surface provides the electrical contact. LCDs with multiplexed addres-

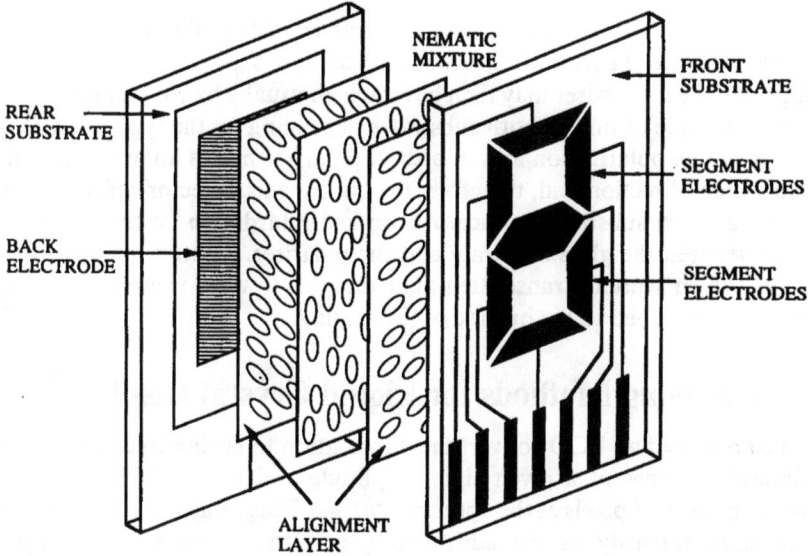

Figure 2.13 *Schematic representation of a simple, low-information-content, alpha-numeric TN-LCD with direct addressing and segmented electrodes.*

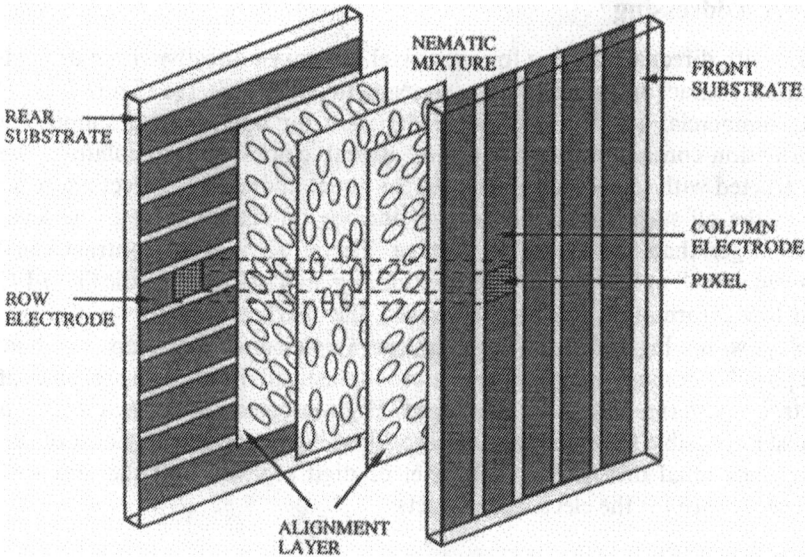

Figure 2.14 *Schematic representation of the pixels, made up of a pattern of orthogonal rows and columns of electrodes, of an LCD with multiplexed addressing.[12]*

sing require a series of uniform rows of electrodes on one surface and an equivalent, sometimes identical, pattern of electrode columns on the other surface. The rows and columns are arranged orthogonal to each other to form a pattern of rectangular or square pixels. LCDs with active matrix addressing use one silicon substrate covered with a regular array of thin film transistors and a back electrode on the second substrate surface.

A plastic sheet polariser may then be attached, usually by contact bonding, to the outer surface of one or both substrates depending on the type of LCD, see Chapter 3. The polarisation axis of a sheet polariser makes an angle, α or β to the alignment direction and, therefore, the optical axis (director) of the nematic mixture, at each substrate surface. An optical retardation layer may also be present between the glass substrate and the polariser. Another passive optical element, which scatters transmitted light through a wider viewing angle cone may also be attached to the outside of the display.

5   Addressing Methods for Liquid Crystal Displays[12]

The electronics of an LCD convert the information to be displayed into a series of applied voltages of a given sign, amplitude and pulse width in order to activate the desired pixels at the appropriate time. This results in the modulation of the light intensity at the addressed pixels, thus creating the image or information to be displayed. The main methods by which this can be achieved are described below.

Direct Addressing

LCDs with direct addressing for each pixel are driven directly with a dedicated electrical contact and a driver for each segment of the digit, see Figure 2.13. The first commercial LCDs, which were designed for operation in simple, low-information-content displays such as digital clocks and calculators, were constructed with segmented electrodes for direct addressing. Direct addressing allows the off-state voltage to be zero and the on-state voltage to be several times larger than the threshold voltage. Therefore, a good contrast can be attained as well as low power consumption, *e.g.* in a twisted nematic (TN)-LCD with low information content. However, the market demand for flat panel displays with a higher information content, *i.e.* personal organisers, notebooks and portable computers, meant that more sophisticated addressing schemes and electrode patterns had to be developed. High-information-content LCDs are almost impossible to realise with direct addressing due to the high cost of using many individual drivers for each pixel or digit segment and the absence of sufficient space for the electrical contacts.

Multiplex Addressing[12,82]

Multiplex addressing with M electrode columns and N electrode rows allows M × N pixels to be created driven by M + N connections made at the end of each row and column, see Figure 2.14. This type of addressing allows large-area, high-information-content LCDs to be produced with acceptable contrast and viewing angles, see Chapter 3.

In an LCD with multiplex addressing, parallel rows and columns of electrodes arranged perpendicular to one another create a matrix of rows and columns. A scan or select voltage pulse (V_S) is applied to each row sequentially, whereas the columns are addressed by data voltage pulses (V_D). These incorporate the information to be displayed. When the amplitude of the resultant voltage of the row and column voltages in phase is larger than the threshold voltage, the nematic director is reoriented by the electric field and the pixel addressed is in the on-state. If a non-select voltage (V_{NS}) below the threshold voltage is applied to the addressed pixel, the elastic forces restore the original alignment configuration of the nematic director in the off-state. This is the case for the commercial LCDs using nematic mixtures described in Chapter 3, although other LCD types using nematic liquid crystals may use voltage pulses of opposite sign to switch a pixel on and off. Root mean square (RMS) voltage is usually applied in most commercial LCDs, since the reorientation of the nematic director is a result of the coupling between the electric field with the induced polarisation of the nematic medium, which is proportional to the square of the applied electric field.

The number of addressable lines in a multiplexed LCD with acceptable contrast and viewing angle dependency is limited, although to a much lesser extent than using direct addressing. The maximum number of addressable lines (N) is given by (see Figure 2.15):[82]

$$\frac{V_{NS}}{V_S} = \sqrt{\frac{\sqrt{N-1}}{\sqrt{N+1}}} \tag{11}$$

Therefore, the more lines there are to be addressed, the smaller the difference between the select and non-select voltages becomes, *i.e.* there is only a difference of about 11% between V_S and V_{NS} for 64 addressable lines. An undesirable consequence of this small voltage difference is the inadvertent activation of adjacent pixels (cross-talk), which reduces the contrast. In order to address a

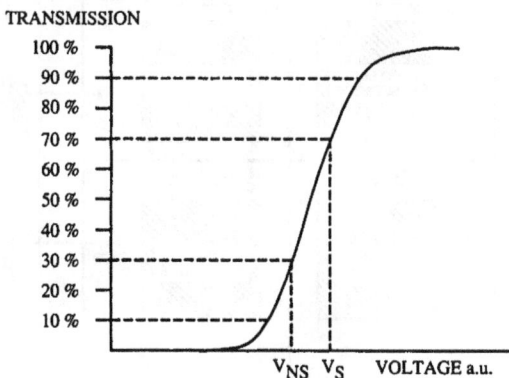

Figure 2.15 *Schematic representation of a typical plot of the transmission versus applied voltage for an LCD with multiplexed addressing. V_S and V_{NS} are the select and the non-select voltages, respectively.*[12]

large number of lines with good contrast, the electro-optical response curve, *i.e.*
transmission *versus* voltage, should exhibit as steep a slope as possible, see
Figure 2.15, although this inadvertently reduces the degree of grey-scale. This
complicates the generation of full colour, since grey-scale allows the intensity of
light transmitted to be controlled by modulating the amplitude and pulse width
of the applied voltage, see Figure 2.15. Consequently, LCDs with multiplexed
addressing are only capable of displaying a limited amount of information with
good contrast over a relatively wide range of viewing angles, unless the electro-
optical transmission curve is very steep, *e.g.* for STN-LCDs, see Chapter 3.
LCDs without a steep electro-optical transmission curve, such as TN-LCDs, see
Chapter 3, cannot display high information-content with good contrast and
wide viewing angles.

Active Matrix Addressing

LCDs with high-information-content are presently realised commercially by
using versions of passively addressed TN-LCDs and STN-LCDs with multiplexed addressing, or ECB-LCDs, TN-LCDs or in-plane switching (IPS)-LCDs
with active matrix addressing, which are all discussed in Chapter 3.[83–87] Most
commercial LCDs with active matrix addressing use a discrete thin-film
transistor (TFT)[84] or much less frequently a diode[85] at each pixel of an array
of many tens of thousands or even hundreds of thousands of pixels fabricated
on an amorphous silicon substrate,[86] see Figure 2.16 for the most common
commercial LCD with active matrix addressing, *i.e.* a TN-TFT-LCD, see

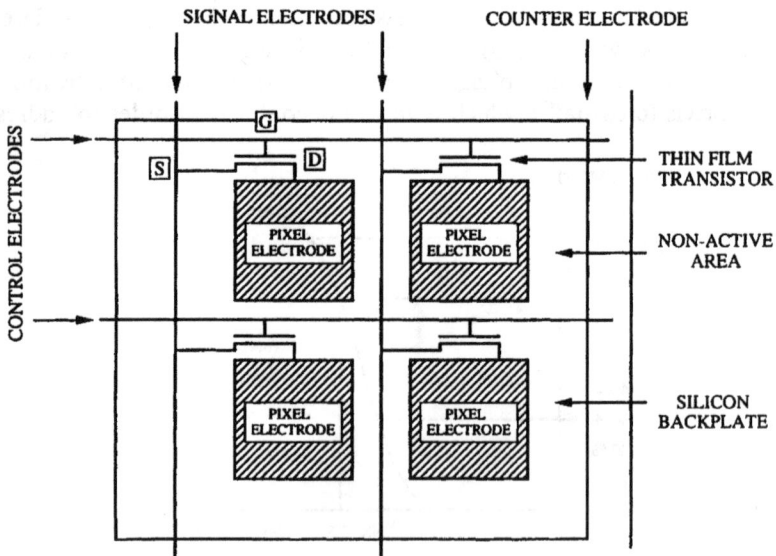

Figure 2.16 *Schematic representation of a high-information-content LCD with active*
matrix addressing provided by thin film transistors (TFTs).[12]

Chapter 3. Active matrix addressing using cadmium selenide (CdSe) thin film-transistors was developed in the early 1960s to address several kinds of display devices.[88] High-information-content, large-area LCDs with active matrix addressing generally exhibit high-contrast, wide-viewing angles, absence of cross-talk, fast response times and adequate grey-scale for full colour. A compromise between brightness and high power consumption must be made due to the low luminosity of such LCDs with two polarisers. The cost of manufacture of LCDs with active matrix addressing is high due to the presence of the active substrate. However, the cost is falling steadily with improved manufacturing methods and machinery and pixel repair techniques.

LCDs with active matrix addressing are switched on by a voltage pulse at the thin film transistor or diode. The charge at the pixel, which corresponds to a capacitor, should remain constant until the pixel is addressed again in the next frame. Leakage of charge away from the pixel into the nematic mixture reduces the effective voltage at the pixel. This results in lower contrast. This effect can be counteracted by using very pure nematic materials with a very high resistivity value, see Chapter 3.

6. Organic Polymer Alignment Layers

The LCDs described in Chapter 3 are all based on field effects. Either the on-state or the off-state are stabilised or regenerated in the absence of an applied field by an alignment layer of some kind covering the inner cell surface in some way or another.[89–91] A nematic phase deposited on an aligned substrate spontaneously develops an equilibrium, or easy orientation, at the interface between the alignment layer and the nematic material. The easy orientation is characterised by the polar angle θ_0 and the azimuthal angle ϕ_0, see Figure 2.12. The corresponding polar and azimuthal anchoring energies, $F_a(\theta)$ and $F_a(\phi)$, respectively, represent the energy required to deform the director from θ_0 and ϕ_0, respectively. These anchoring energies can be expressed in the Rapini–Papoular approximations

$$F_p(\theta) = W_\theta \sin^2(\theta - \theta_0) \tag{12}$$
$$F_a(\phi) = W_\phi \sin^2(\phi - \phi_0) \tag{13}$$

where W_θ is the polar part of the anchoring energy and W_ϕ is the azimuthal part of the anchoring energy with $\phi_0(\theta_0)$ fixed. The polar anchoring energy may be an order of magnitude larger than the azimuthal anchoring energy, *e.g.* $W_\theta > 10^{-3} \, \mathrm{J \, m^{-2}}$ and $W_\phi > 10^{-4} \, \mathrm{J \, m^{-2}}$ for strongly rubbed polyimide.[92] The interaction between the substrate and the liquid crystalline phase is described phenomenologically by the ratio of the surface tension of the surface (γ_S) and the surface tension of the liquid crystal (γ_{LC}). The alignment is homeotropic if the intermolecular forces in the bulk of the liquid crystalline phase are larger than the forces between the surface and the liquid crystal ($\gamma_S < \gamma_{LC}$). Homogeneous alignment occurs when the surface anchoring forces are greater. The nature of the interaction between the surface and the liquid crystal molecules is complex and includes dispersive (Van der Waals), polar, steric, order-electric and ionic terms. The topology of the surface may also be an important factor as indicated by the alignment of the nematic director parallel to the grooves of gratings in order to minimise the elastic strain.[93] A unique azimuthal orientation usually requires surface treatment of the alignment layer, for example, by mechanical rubbing or illumination with polarised ultra violet (UV) light.

A self-assembled monolayer (SAM) of long aliphatic chains fixed to the substrate surface through reactive siloxane endgroups is the common method used to generate homeotropic alignment.[94] Aliphatic SAMs usually generate homeotropic alignment, perhaps with a small pretilt angle measured from the normal to the plane of the cell ($\theta_0 \approx 1°$), due to the low value of the surface tension attributable to the planar surface of apolar methyl groups in a terminal position. However, most LCD types require homogeneous alignment with the director inclined at a small angle to the plane of the cell. Originally inorganic oxides such as SiO_2 were deposited from an oblique angle under high vacuum to generate homogeneous alignment.[95] Nematic liquid crystals wet these inorganic surfaces producing high ($\approx 10°$) pretilt angles, θ, where $\theta = 90° - \theta_0$. However, the production technique to produce these alignment layers is intrinsically expensive and not compatible with continuous production processes. The most common materials used over the last two decades at least to create an orientation effect are polyimide derivatives with a low value for the surface tension, such as those (**24–29**) shown in Table 2.4.[96–100] A soluble precursor polyamide is first deposited and then cured at high temperature to form a uniform layer of insoluble and intractable polyimide with very high glass transition temperatures and decomposition temperatures. In commercial LCD fabrication these layers are rubbed (buffed) mechanically in one direction to create a unique orientation direction in the azimuthal and zenithal planes.[90,91] The buffing of organic polymers is the standard alignment technique for homogeneous alignment with a defined pretilt angle in commercial LCDs.[96–100]

Mechanical rubbing can cause pixel damage and generate static electricity leading to dielectric breakdown or higher conductivity of the nematic layer. This is especially important for LCDs with active matrix addressing, where the presence of even very small amounts of dust or charged particles can lead to lower production yields and higher manufacturing costs. Non-contact alignment layers use plane polarised UV light to generate a surface anisotropy.[91] This, in turn, can lead to a uniform alignment of the nematic director. Therefore, they have considerable potential to replace buffed polymer orientation layers especially for LCDs with active matrix addressing. A diverse range of related non-contact alignment methods making use of the same effect, *i.e.* the *cis–trans* photoisomerisation of azo-dyes, see Figure 2.17, to produce an orientation effect of the nematic director by co-operative effects, has been developed.[101–109] The azo dye may be located in the orientation layer,[101] the nematic mixture,[102,103] or covalently bonded to polymers deposited on the substrate surface[104–106] or in monolayers deposited by Langmuir–Blodgett techniques.[107–109] Non-contact alignment layers using azo-chromophores require the use of laser light for relatively long exposure times to induce the

Table 2.4 *Polyimides (24–29) used as alignment layers, especially as non-contact alignment layers, for LCDs.*

Molecular structure	Ref
	96
	96
	97
	98

(continued)

desired isomerisation of the dye molecules, which are present in high concentration, *e.g.* 2 wt%. This may result in an undesirable colouration of the display. Unfortunately, the thermal and UV stability of the orientation layers may be insufficient for commercial LCDs.

Poly(vinylcinnamate) (PVC) films exposed to linearly polarised ultraviolet

Table 2.4 *continued*

Molecular structure	Ref

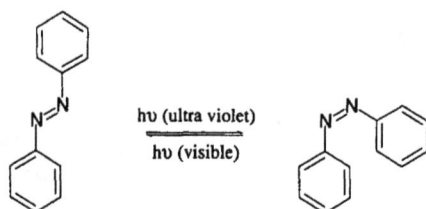

28   99

29   100

Figure 2.17 *Cis (Z) and trans (E) isomeric forms of azobenzene.*

light can also generate homogeneous alignment of the nematic director.[91,110–116] The resultant orientation is perpendicular to the polarisation direction of the plane polarised UV light. The alignment is caused by an anisotropic depletion of the cinnamate sidechains as a consequence of a 2 + 2 cycloaddition reaction, although *cis–trans* isomerisation of the carbon–carbon double bond may also be present at low fluences, see Figure 2.18.[112] A low azimuthal surface anchoring energy ($\approx 4 \times 10^{-6}$ J m^{-2}) has been found for PVC itself.[113] However, stronger anchoring of the nematic director is found when derivatives of PVC are used.[114–116] Small pretilt angles ($\theta \approx 0.3°$) normal to the surface, which are required to avoid areas of reverse tilt (see Chapter 3) have been generated by a double exposure of the PVC layer.[114] However, larger pretilt angles have been achieved by exposure at non-normal incidence of a coumarin side-chain polymer,[117,118] see Figure 2.19. The alignment direction is parallel to the polarisation direction of the UV light. LCDs with improved viewing-angle-dependence of the contrast can be realised using this technique to create sub-

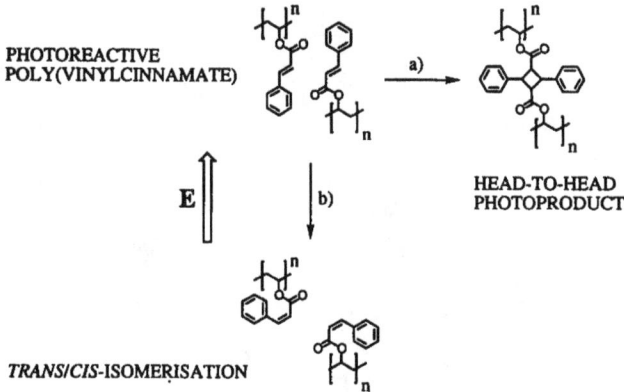

Figure 2.18 *Photochemical reactions of PVC on irradiation of UV light: (a) 2 + 2 cycladdition and (b) trans/cis-(E/Z)-isomerisation. E indicates the polarisation direction of the incident beam.*

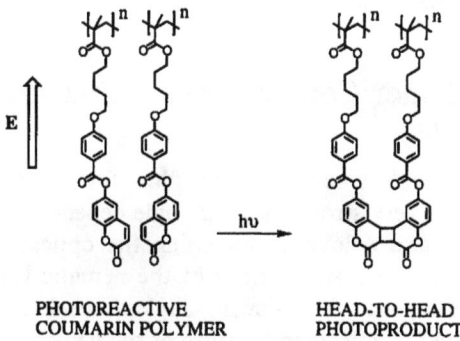

Figure 2.19 *Schematic representation of the anisotropic crosslinking of a photoreactive coumarin polymer by the action of polarised UV light to produce an anisotropic network as a non-contact alignment layer. E indicates the polarisation direction of the incident beam.[117–123]*

domain pixels, each with a different microscopic alignment direction.[119,120] Anchoring energies comparable to those achieved using standard rubbed polyimide have been determined for non-contact alignment layers. [121–126]

The anisotropic photodegradation of polyimide layers, such as those polyimides shown in Table 2.4, using plane polarised UV light can also be used to induce uniform alignment of the nematic director by a non-contact method.[125,126] However, the ablation process proceeds by breaking chemical bonds and removing the debris thermally. This generates substantial amounts of chemical decomposition products and a rough surface. These can pose a serious problem for LCDs with active matrix addressing, if the debris is not completely removed from the substrate surface.

TRANSMITTED
LIGHT WAVE FRONT

ZENITHAL AXIS

AZIMUTHAL AXIS

INCIDENT
LIGHT WAVE FRONT

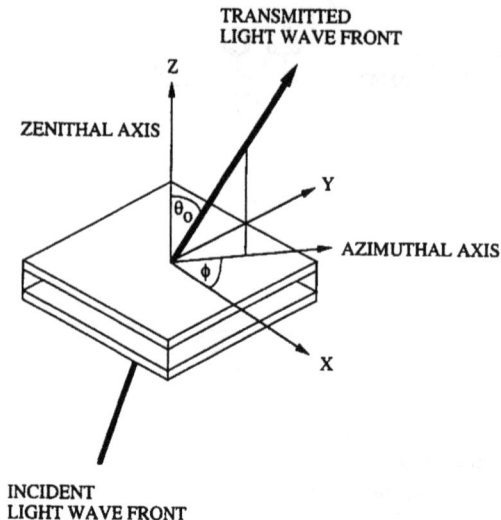

Figure 2.20 *Schematic representation of viewing angles, θ_0 and ϕ, in the zenithal and azimuthal planes for light traversing an LCD.*[12]

7 Organic Polymer Compensation Films for Liquid Crystal Displays[127-141]

Optical retarders are used as compensation films for some types of LCD in order to correct for the very strong viewing angle dependence of the contrast, luminosity, grey-scale and colour by matching the optical symmetry of the compensation layer to the optical pattern of the nematic layer, but with the opposite sign of birefringence. The viewing-angle dependence of these optical properties is caused by the anisotropic nature of the nematic layer in the LCD and polarisers.[127] These optical properties are optimised in commercial LCDs for viewing directly from above. Therefore, the optical performance of an LCD usually degrades as the angle of view increases from the normal to the plane of the cell, see Figure 2.20. This viewing angle dependence is also asymmetric, *i.e.* it depends on the direction of view and not just the angle of view. These problems are particularly acute for LCDs based on wave-guiding, such as TN-LCDs, and LCDs based on interference effects, such as ECB-LCDs and STN-LCDs (see Chapter 3). The optical properties of LCDs with the optical axis in the plane of the cell, such as IPS-LCDs, are much less viewing-angle dependent (see Chapter 3).

There is an almost infinite potential to combine different kinds of compensation films in many different ways to achieve a given effect, *e.g.* compensation layers of negative birefringence with the optical axes in the plane of the cell situated on either side of a TN-LCD improve the viewing angle dependence in the horizontal plane, but not in the vertical plane.[128-141] However, if the optical axis of the compensation layer is tilted with respect to the plane of the cell, then superior optical properties in the vertical direction are also achieved. Compen-

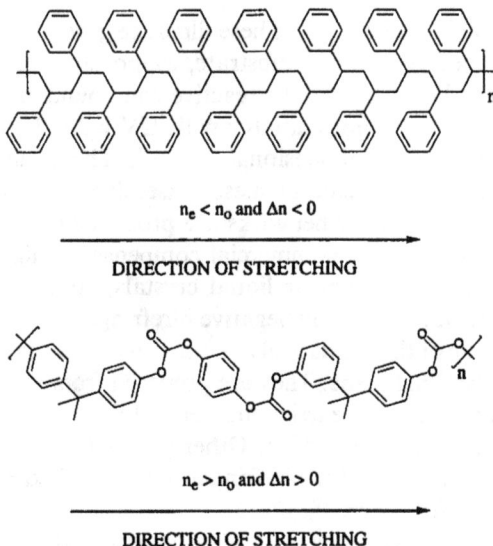

$n_e < n_o$ and $\Delta n < 0$

DIRECTION OF STRETCHING

$n_e > n_o$ and $\Delta n > 0$

DIRECTION OF STRETCHING

Figure 2.21 *Schematic representation of the linear structure of a stretched polycarbonate and a stretched polystyrene component of an optical retardation sheet.*

sation films may be organic layers of positive or negative birefringence, whose optical axis makes an angle to the plane of the cell. This angle may be fixed or increase in a splayed configuration of the nematic director. The optical axes of individual layers may be parallel to the nematic director at the same substrate surface or the major axis of absorption of the polariser attached to this substrate, or may make some angle to either one or both of them. A number of stacked compensation layers may be fixed either above and/or below the cell situated between the polariser and the substrate. Computer programs, such as the Jones or Berreman programs,[139] are often used to design compensation films by predicting the effect of a compensation layer of a given birefringence, thickness and orientation of the optical axis on the optics of the LCD in question.

Simple compensation films consist of oriented films of mainchain polymers incorporating polarisable aromatic units, *e.g.* a polystyrene or polycarbonate polymer, see Figure 2.21. The films are mechanically stretched to align the polymer chains parallel to each other in the direction of mechanical stress.[127] The stretching process creates a polarisation anisotropy, since the polarisability parallel to the long polymer axis is greater than that orthogonal to it for the polycarbonate. Therefore, the retardation film is of positive birefringence, since n_e is greater than n_o (see Equation 7). A stretched styrene polymer sheet[128] exhibits a negative birefringence, since the refractive index orthogonal to the oriented polymer chain will be larger than that parallel to the chain, since n_e is less than n_o (see Figure 2.21).

More sophisticated optical compensation layers can be produced as anisotropic networks from reactive mesogens (liquid crystals) in a macroscopically aligned liquid crystalline state.[129–141] These films are generally prepared by spin casting a reactive mesogen onto a substrate, evaporating off the solvent and then polymerising and crosslinking the reactive compound in the nematic state using a small amount of a photoinitiator with UV light. This process can be used to produce a solid three-dimensional network. The nematic liquid state is usually preferred to any of the smectic mesophases due to the lower viscosity of the former. However, columnar networks are produced commercially, despite their very high viscosity. Some commercial compensation films, *e.g.* Fuji WV Film Wide View A, using columnar liquid crystals, are a hybrid of multiple compensation layers of positive and negative birefringence with the optical axis in the azimuthal plane of the cell as well as configured with the optical axis tilted and splayed in the zenithal plane. They are prepared from photopolymerisable columnar liquid crystals on a carrier film of cellulose triacetate, which also contributes to the compensation effect. Other products incorporate a nematic anisotropic network with positive birefringence and a tilted optical axis[135] or holographic form birefringence.[136]

There are many other processing methods which can reduce the viewing angle dependence of the optical properties of LCDs, *e.g.* pixel-divided cells[140] and multidomain cells.[141] However, these are beyond the scope of this monograph and will not be discussed further. Optical retardation sheets are often preferred by LCD manufacturers and systems integrators of FPDs, because commercially available compensation films can simply be laminated to the LCD using contact bonding.

Optically anisotropic non-contact alignment layers could also be used to generate photo-patterned, high-resolution optical retarders, but within the cell rather than as an external film.[122,123] This could offer practical advantages, since there would be no need to fix compensation layers to the LCD by contact bonding. However, the birefringence of such alignment layers is generally low. Consequently, very thick films would have to used in order to obtain the desired retardation. However, they could be used in hybrid compensation layers in combination with anisotropic networks of high birefringence formed from reactive mesogens.[77,119] Macroscopically oriented films can be formed by orientation of the reactive mesogen in the nematic state on the photoalignment layer, which is then cross-linked to form a patterned optical retarder as an anisotropic network.

8   References

1 D. Dunmur and K. Toriyama, in 'Physical Properties of Liquid Crystals', Eds. D. Demus, J. W. Goodby, G. W. Gray, H.-W. Spiess and V. Vill, Wiley-VCH Verlagsgesellschaft GmbH, Weinheim, Germany, 1999, p. 85.
2 V. Fréederieks and V. Zolina, *Trans. Faraday Soc.*, 1933, **29**, 919.
3 V. Zwetkoff, *Acta Physico. Chem. USSR*, 1937, **6**, 885.
4 P. Chatelain, *Bull. Soc. Fr. Minéral. Cristallogr.*, 1943, **66**, 105.
5 R. Williams, *J. Chem. Phys.*, 1963, **39**, 384.
6 W. Helfrich, *Mol. Cryst. Liq. Cryst.*, 1973, **21**, 187.
7 M. Schadt and W. Helfrich, *Appl. Phys. Lett.*, 1971, **18**, 127.

8 G. Baur, *Mol. Cryst. Liq. Cryst.*, 1981, **63**, 45.
9 T. J. Scheffer and J. Nehring, *Appl. Phys. Lett.*, 1984, **45**, 1021.
10 C. Waters, V. Brimmel and E. P. Raynes, *Proc. Japan Display '83*, 1983, 396.
11 L. M. Blinov, in 'Handbook of Liquid Crystal Research', Eds. P. J. Collings and J. S. Patel, Oxford University Press, Oxford, UK, 1997.
12 T. J. Scheffer and J. Nehring, *Ann. Rev. Mater. Sci.*, 1997, **27**, 555.
13 M. Schadt, *Ann. Rev. Mater. Sci.*, 1997, **27**, 305.
14 V. G. Chigrinov, in 'The Physics of Liquid Crystal Display Devices', Springer, New York, USA, 1998.
15 I. C. Sage, in 'Handbook of Liquid Crystals', Eds. D. Demus, J. W. Goodby, G. W. Gray, H.-W. Spiess and V. Vill, Wiley-VCH Verlagsgesellschaft GmbH, Weinheim, Germany, 1998, Vol. 1, p. 731.
16 L. M. Blinov, in 'Handbook of Liquid Crystals' Eds. D. Demus, J. W. Goodby, G. W. Gray, H.-W. Spiess and V. Vill, Wiley-VCH Verlagsgesellschaft GmbH, Weinheim, Germany, 1998, Vol. 1, p. 477.
17 L. M. Blinov and V. G. Chigrinov, in 'Electro-optic Effects in Liquid Crystal Materials', Springer, New York, USA, 1994.
18 E. P. Raynes, in 'The Optics of Thermotropic Liquid Crystals', Eds. S. Elston and R. Sambles, Taylor and Francis, London, UK, 1998, p. 289.
19 S. Garoff and R. B. Meyer, *Phys. Rev. Lett.*, 1997, **38**, 848.
20 R. B. Meyer, *Mol. Cryst. Liq. Cryst.*, 1977, **40**, 33.
21 S. T. Lagerwall, in 'Handbook of Liquid Crystals', Eds. D. Demus, J. W. Goodby, G. W. Gray, H.-W. Spiess and V. Vill, Wiley-VCH Verlagsgesellschaft GmbH, Weinheim, Germany, 1998, Vol. 2B, p. 515.
22 A. Fukuda, *J. Mater. Chem.*, 1994, **4**, 997.
23 V. Percec and C. Pugh, in 'Side Chain Liquid Crystalline Polymers', Ed. C. B. McArdle, Blackie, Glasgow and London, UK, 1989, p. 30.
24 J. C. Dubois, P. le Barny, M. Mauzac and C. Noel, in 'Handbook of Liquid Crystals', Eds. D. Demus, J. W. Goodby, G. W. Gray, H.-W. Spiess and V. Vill, Wiley-VCH Verlagsgesellschaft GmbH, Weinheim, Germany, 1998, Vol. 3, p. 207.
25 'Handbook of Liquid Crystals', Eds. D. Demus, J. W. Goodby, G. W. Gray, H.-W. Spiess and V. Vill, Wiley-VCH Verlagsgesellschaft GmbH, Weinheim, Germany, 1997.
26 V. Vill, *Liq. Cryst.*, 1998, **24**, 21.
27 V. Vill, 'LiqCryst. 1 – Database of Liquid Crystalline Compounds for Personal Computers', LCI Publisher, Hamburg, Germany, 1996.
28 D. Blunk, K. Praefcke and V. Vill, in 'Handbook of Liquid Crystals', Eds. D. Demus, J. W. Goodby, G. W. Gray, H.-W. Spiess and V. Vill, Wiley-VCH Verlagsgesellschaft GmbH, Weinheim, Germany, 1998, Vol. 3, p. 305.
29 C. Fairhurst, S. Fuller, J. Gray, M. C. Holmes and G. J. T. Tiddy, in 'Handbook of Liquid Crystals', Eds. D. Demus, J. W. Goodby, G. W. Gray, H.-W. Spiess and V. Vill, Wiley-VCH Verlagsgesellschaft GmbH, Weinheim, Germany, 1998, Vol. 3, p. 341.
30 F. Reinitzer, *Monatsch. Wiener Chem. Ges.*, 1888, **9**, 421.
31 O. Lehmann, *Z. Phys. Chem., Leipzig*, 1889, **4**, 462.
32 D. Vorländer, in 'Kristalline Flüßigkeiten und Flüßige Kristalle', Engelmann, Leipzig, Germany, 1905.
33 D. Demus, in 'Handbook of Liquid Crystals', Eds. D. Demus, J. W. Goodby, G. W. Gray, H.-W. Spiess and V. Vill, Wiley-VCH Verlagsgesellschaft GmbH, Weinheim, Germany, 1998, Vol. 1, 134.
34 H. Sackmann, *Liq. Cryst.*, 1989, **5**, 43.
35 G. W. Gray, in 'Molecular Structure and the Properties of Liquid Crystals', Academic Press, London, UK, 1962.
36 G. W. Gray, in 'Advances in Liquid Crystals', Academic Press, New York, USA, 1976, Vol. 2, p. 1.

37 G. W. Gray, *Phil. Trans. R. Soc. Lond.*, 1983, **A309**, 77; 1990, **A330**, 73.
38 R. E. Rondeau, M. A. Berwick, R. N. Steppel and M. P. Servé, *J. Am. Chem. Soc.*, 1972, **94**, 1096.
39 R. Steinsträsser, *Z. Naturforsch.*, 1972, **27b**, 774.
40 J. A. Castellano, J. E. Goldmacher, L. A. Barton and J. S. Kane, *J. Org. Chem.*, 1968, **33**, 3501.
41 R. Steinsträsser and L. Pohl, *Z. Naturforsch.*, 1971, **26b**, 577.
42 M. Rosenberg and R. A. Champa, *Mol. Cryst. Liq. Cryst.*, 1970, **11**, 191.
43 H. Kelker and B. Scheurle, *Angew. Chem. Int. Ed. Eng.*, 1969, **8**, 884.
44 H. Zocher, *Trans. Faraday Soc.*, 1933, **29**, 931 and 945.
45 C. W. Oseen, *Trans. Faraday Soc.*, 1933, **29**, 883.
46 F. C. Frank, *Discuss. Faraday Soc.*, 1958, **25**, 19.
47 J. L. Eriksen, *Art. Rat. Mech. Anal.*, 1962, **9**, 371.
48 J. L. Eriksen, *Trans. Soc. Rheol.*, 1961, **5**, 23.
49 F. M. Leslie, *Art. Rat. Mech. Anal.*, 1968, **28**, 265.
50 F. M. Leslie, in 'Physical Properties of Liquid Crystals', Eds. D. Demus, J. W. Goodby, G. W. Gray, H.-W. Spiess and V. Vill, Wiley-VCH Verlagsgesellschaft GmbH, Weinheim, Germany, 1999, p. 25.
51 W. Maier and A. Saupe, *Z. Naturforsch.*, 1958, **13a**, 564; 1959, **14a**, 882; 1960, **15a**, 287.
52 R. Alben, *Mol. Cryst. Liq. Cryst.*, 1971, **13**, 193.
53 P. J. Flory, and G. Ronca, *Mol. Cryst. Liq. Cryst.*, 1979, **54**, 311.
54 M. A. Cotter, *Phil. Trans. R. Soc., London*, 1983, **A309**, 127.
55 M. A. Osipov, in 'Physical Properties of Liquid Crystals', Eds. D. Demus, J. W. Goodby, G. W. Gray, H.-W. Spiess and V. Vill, Wiley-VCH Verlagsgesellschaft GmbH, Weinheim, Germany, 1999, p. 40.
56 P. G. de Gennes, in 'The Physics of Liquid Crystals', Oxford University Press, UK 1974.
57 P. G. de Gennes and J. Prost, in 'The Physics of Liquid Crystals', Oxford University Press, Oxford, UK, 1993.
58 U. Finkenzeller, *Kontakte (Darmstadt)*, 1988, **2**, 7.
59 G. W. Gray, K. J. Harrison and J. A. Nash, *Electron. Lett.*, 1973, **9**, 130.
60 P. Chatelain, *Bull. Soc. Franc. Mineral Crist.*, 1954, **77**, 353.
61 M. Miezowicz, *Nature*, 1946, **158**, 27.
62 G. H. Heilmeier and L. A. Zanoni, *Appl. Phys. Lett.*, 1968, **13**, 91.
63 G. H. Heilmeier, L. A. Zanoni and L. A. Barton, *Proc. IEEE*, 1968, **56**, 1162.
64 P. A. Penz, *Phys. Rev. Lett.*, 1970, **24**, 1405.
65 E. F. Carr, *Mol. Cryst. Liq. Cryst.*, 1969, **7**, 253.
66 W. Helfrich, *J. Chem. Phys.*, 1969, **51**, 4092.
67 R. B. Meyer, *Phys. Rev. Lett.*, 1969, **22**, 918.
68 D. Schmidt, M. Schadt and W. Helfrich, *Z. Naturforsch.*, 1972, **27a**, 277.
69 J. S. Patel and R. B. Meyer, *Phys. Rev. Lett.*, 1987, **58**, 1538.
70 I. Dozov, M. Nobili and G. Durand, *Appl. Phys. Lett.*, 1997, **70**, 1179.
71 G. P. Bryan-Brown, C. V. Brown, J. C. Jones, E. L. Wood, I. C. Sage, P. Brett and J. Rudin, *Proc. SID '97 Digest*, 1997, 37.
72 J. L. Ferguson, *SID Digest Tech. Papers.*, 1985, **16**, 68.
73 D. Coates, *J. Mater. Chem.*, 1995, **5**, 2063.
74 G. P. Crawford, J. W. Doane and S. Zumer, in 'Handbook of Liquid Crystal Research', Eds. P. J. Collings and J. S. Patel, Oxford University Press, Oxford, UK, 1997, p. 347.
75 D. J. Broer, R. A. M. Hikmet and G. Challa, *Makromol. Chem.*, 1989, **190**, 3201.
76 J. Lub, D. J. Broer, R. A. M. Hikmet and K. G. J. Nierop, *Liq. Cryst.*, 1995, **18**, 319.
77 S. M. Kelly, *J. Mater. Chem.*, 1995, **5**, 2047.
78 D.-K. Yang and J. W. Doane, *SID '92 Digest*, 1992, 759.
79 U. Behrens and H.-S. Kitzerow, *Polym. Adv. Tech.*, 1994, **5**, 433.

80 R. A. M. Hikmet, H. M. J. Boots and M. Michielson, *Liq. Cryst.*, 1995, **19**, 65.
81 R. A. M. Hikmet, *Adv. Mater.*, 1992, **4**, 679.
82 P. M. Alt and P. Pleshko, *IEEE Trans. Electron. Devices*, 1974, **21**, 146.
83 S. Kobayashi, H. Hori and Y. Tanaka, in 'Handbook of Liquid Crystal Research', Eds. P. J. Collings and J. S. Patel, Oxford University Press, Oxford, UK, 1997.
84 B. J. Lechner, *Proc. IEEE*, 1971, **59**, 1566.
85 D. E. Castlebury, *IEEE Trans. Elec. Dev.*, 1979, **ED-26**, 1123.
86 D. Dixon, *IEEE Trans. Elec. Dev.*, 1973, **ED-20**, 995.
87 E. Kaneko in, 'Handbook of Liquid Crystals', Eds. D. Demus, J. W. Goodby, G. W. Gray, H.-W. Spiess and V. Vill, Wiley-VCH Verlagsgesellschaft GmbH, Weinheim, Germany, Vol. 2A, 1998, 230.
88 T. P. Brody, J. T. Asars and G. D. Dixon, *IEEE Trans. Elec. Dev.*, 1973, **ED-20**, 995.
89 J. Cognard, *Mol. Cryst. Liq. Cryst.*, 1982, **51**, 1.
90 D. Berreman, *Phys. Rev.*, 1972, **28**, 1683.
91 M. O'Neill and S. M. Kelly, *J. Phys. D: Appl. Phys.*, 2000, **33**, R67.
92 M. G. Tomlin, *J. Opt. Technol.*, 1997, **64**, 458
93 D. W. Berreman, *Mol. Cryst. Liq. Cryst.*, 1973, **23**, 215.
94 F. J. Kahn, *Appl. Phys. Lett.*, 1973, **22**, 386.
95 L. Janning, *Appl. Phys. Lett.*, 1972, **21**, 173.
96 M. Nishikawi, K. Tamas and J. L. West, *Jpn. J. Appl. Phys.*, 1999, **38**, L334.
97 J-H. Kim, B. R. Acharya and S. Kumar, *Appl. Phys. Lett.*, 1998, **73**, 3372.
98 M. Nishikawi, B. Taheri and J. L. West, *Appl. Phys. Lett.*, 1998, **72**, 2403.
99 B. Park, Y. Jung, H. Choi, H. Hwang, M. Kakimoto and H. Takezoe, *Jpn. J. Appl. Phys.*, *1*, 1998, **37**, 5663.
100 Y. Wang, C. Xu, A. Kanazawa, T. Shiono, T. Ikeda, Y. Matsuki and Y. Takeuchi, *J. Appl. Phys.*, 1998, **84**, 181.
101 W. M. Gibbons, P. J. Shannon, S.-T. Sun and B. J. Swetlin, *Nature*, 1991, **351**, 49.
102 S.-T. Sun, W. M. Gibbons and P. J. Shannon, *Liq. Cryst.*, 1992, **12**, 869.
103 W. M. Gibbons, P. J. Shannon and S.-T. Sun, *Mol. Cryst. Liq. Cryst.*, 1994, **251**, 191.
104 K. Ichimura, N. Hamada, S. Kato and I. Shinohara, *J. Polym. Sci., Polym. Chem. Ed.*, 1983, **21**, 1551.
105 Y. Imura, J. Kusano, S. Kobayashi and Y. Aoyagi, *Jpn. J. Appl. Phys.*, 1993, **32**, L93.
106 K. Ichimura, Y. Suzuki, T. Seki, A. Hosoki and N. Aoki, *Langmuir*, 1988, **4**, 1214.
107 K. Ichimura, in 'Photochemical Processes in Organized Molecular Systems', Ed. K. Honda, Elsevier, Amsterdam, The Netherlands, 1991, p. 343.
108 K. Ichimura, Y. Hayashi, Y. Kawanishi, T. Seki, T. Tamaki and N. Ishizuki, *Langmuir*, 1993, **9**, 857.
109 K. Ichimura, Y. Hayashi and N. Ishizuki, *Chem. Lett.*, 1992, 1063.
110 A. Dyadyusha, V. Kozenkov, T. Marusii, Y. Reznikov, V. Reshetnyak, A. Khizhnyak, *Ukr. Fiz. Zh.*, 1991, **36**, 1059.
111 M. Schadt, K. Schmitt, V. Kozinkov and V. G. Chigrinov, *Jpn. J. Appl. Phys.*, 1992, **31**, 2135.
112 K. Ichimura, Y. Akita, K. Akiyama, K. Kudo and Y. Hayashi, *Macromolecules*, 1997, **30**, 903.
113 G. P. Bryan-Brown and I. C. Sage, *Liq. Cryst.*, 1996, **20**, 825.
114 M. Schadt, H. Seiberle, A. Schuster and S. M. Kelly, *Jpn. J. Appl. Phys.*, 1995, **34**, L764.
115 M. Schadt, H. Seiberle, A. Schuster and S. M. Kelly, *Jpn. J. Appl. Phys.*, 1995, **34**, 3240.
116 Y. Imura, S. Kobayashi, T. Hashimoto, T. Sugiyama and K. Katoh, *IEICE Trans. Electron*, 1996, **E79-C**, 1040.
117 R.-P. Herr, F. Herzog and A. Schuster, Patent Application WO 96/10049, 1996.
118 M. Schadt, H. Seiberle and A. Schuster, *Nature*, 1996, **381**, 212.

119 M. Schadt, *Mol. Cryst. Liq. Cryst.*, 1997, **292**, 235.

120 M. Schadt and H. Seiberle, *Proc. SID*, **5/4**, 1997, 367.

121 S. M. Kelly, P. Hindmarsh, G. J. Owen, P. O. Jackson, P. J. Taylor and M. O'Neill, *SID EID '98 Digest*, 1998, 5.1.

122 P. Hindmarsh, G. J. Owen, S. M. Kelly, P. O. Jackson, M. O'Neill and R. Karapinar, *Mol. Cryst. Liq. Cryst.*, 1999, **332**, 439.

123 P. O. Jackson, R. Karapinar, M. O'Neill, P. Hindmarsh, G. J. Owen and S. M. Kelly, *Proc. SPIE*, 1999, **3635**, 48.

124 M. Hasegawa and Y. Taira, *IRDC '94 Digest*, 1994, 213.

125 J. L. West, X. Wang, Y. Ji and J. R. Kelly, *SID '95 Digest*, 1995, 703.

126 C. J. Newsome, M. O'Neill and G. P. Bryan Brown, *Proc. SPIE*, 1999, **3618**, 132.

127 P. J. Bos, *Proc. SID '94*, 1994, 118.

128 Y. Fujimura, *SID Digest*, 1992, 397.

129 S. Kaneko, Y. Hirai and K. Sumiyoshi, *SID Digest '93*, 1993, 265.

130 H. L. Ong, *Japan Display '92*, 1992, 247.

131 J. P. Eblen, W. J. Gunning, J. Beedy, D. Taber, L. Hale, P. Yeh and M. Khoshnevisan, *SID '94 Digest*, 1994, 245.

132 A. Shimizu, *Asia Display '98*, 1998, 207.

133 S. Stallinga, J. M. van den Eerenbeemd and J. A. M. M. van Haaren, *Jpn. J. Appl. Phys.*, 1998, **37**, 560.

134 H. Mori, Y. Itoh, Y. Nisiura, T. Nakamura and Y. Shinagawa, *AM-LCD '96/IDW '96 Digest*, 1996, 189.

135 P. van de Witte, J. van Haaren, J. Tuijtelaars, S. Stallinga and J. Lub, *Jpn. J. Appl. Phys.*, 1999, **38**, 748.

136 C. Joubert and J.-C. Lehureau, *Asia Display '98*, 1998, 1119.

137 R. Brinkley, G. Xu, A. Abileah and J. Vanderploeg, *SID '98 Digest*, 1998, 471.

138 A. Lien, *Appl. Phys. Lett.*, 1990, **57**, 2767.

139 M. Jones and G. Xu, *SID '98 Digest*, 1998, 475.

140 K. R. Sarma, H. Franklin, M. Johnson, K. Frost and A. Bernot, *SID '89 Digest*, 1989, 148.

141 K. H. Yang, *IDRC '91 Digest*, 1991, 68.

CHAPTER 3

Liquid Crystal Displays Using Nematic Liquid Crystals

1 Introduction

The liquid crystals displays (LCDs) currently manufactured in significant volume for the wide variety of instruments that incorporate them in order to display numeric and graphical data all make use of the reorientation of the optical axis of nematic liquid crystals under the influence of an applied electric field. The principles of operation of the most important types of LCDs, which have either been manufactured or are still being manufactured in large numbers, are illustrated in this chapter. Their capacity to display information, their intrinsic limitations as well as those imposed by the nematic materials used are characterised. The development in the design and synthesis of the nematic materials made use of in the first prototypes of each kind of LCD through to the components of complex nematic mixtures used today is also described. The correlation between the molecular structure of these nematic materials and the magnitude of the physical properties required to fulfil the widely differing specifications of the various types of LCD is poorly understood. However, an attempt is made to summarise some of this often empirical knowledge. Representative examples of the most important classes of nematic liquid crystals used in LCDs are described.

2 Dynamic Scattering Mode Liquid Crystal Displays

The first practical electro-optical flat panel display utilising any kind of liquid crystal as the switchable light valve device, was reported by Heilmeier, Zanoni and Barton at the RCA Laboratories in Princeton, New Jersey in 1968.[1] In this particular case the nematic director does not always align itself parallel with the electric field vector for voltages below a certain threshold voltage for dielectric coupling, although this voltage can be very high. The anomalous behaviour is due to the movement under the action of an electric field of a small number of ions within the nematic liquid crystal, which gives rise to electrohydrodynamic instabilities.[2-7] It is the anisotropy of conductivity of liquid crystals which leads

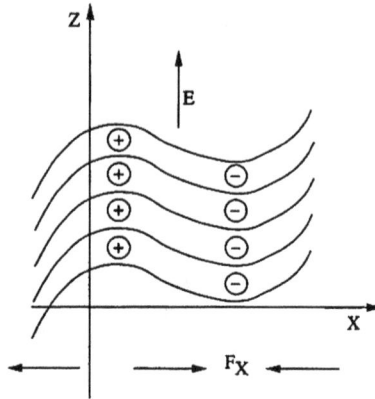

Figure 3.1 *Schematic representation of the turbulence in a nematic medium caused by the movement of ions under the action of an electric field.[2]*

to charge segregation and the generation of space charges. The stirring effect of an electric field on nematic liquid crystals had already been observed in the 1930s.[6] This effect leads to the formation of visible domains, known as Williams domains,[7] at low voltages and light scattering above a higher threshold voltage.[1] The interaction of the electric field with the ions gives rise to transverse fields orthogonal to the applied field if the conductivity is greater parallel to the nematic director than perpendicular to it.[2] This in turn leads to turbulence in the liquid crystalline medium and produces a wavy structure, see Figure 3.1.[2] This results in the movement of areas of liquid crystal in alternate directions to produce Williams domains.[7] The elastic and dielectric torque under steady state conditions counterbalances the shear produced by this electrohydrodynamic effect. At higher fields these stripes are replaced by a homogeneous scattering of incident visible light. Although liquid crystals are generally regarded as dielectrics, *i.e.* non-conductive, insulating organic materials, the conductivity required for the formation of electrohydrodynamic instabilities is relatively low $(10^{-9}\,\Omega^{-1}\,cm^{-1})$. However, the high operating voltages and limited lifetime of many commercial dynamic scattering mode liquid crystal displays (DSM-LCDs) were limiting factors with regard to their initial acceptance in the broader electro-optic market. The advent of competing LCD types with an improved property profile essentially sealed the fate of dynamic scattering mode (DSM)-LCDs.

Display Configuration

The DSM-LCD is a sandwich cell consisting of two parallel glass substrates separated ($6\,\mu m < d > 25\,\mu m$) by spacers (Teflon/Mylar), which usually incorporates a nematic liquid crystal of negative dielectric anisotropy aligned either homeotropically or homogeneously (planar alignment). However, most commercial DSM-LCDs were based on homeotropically aligned cells, see Figure

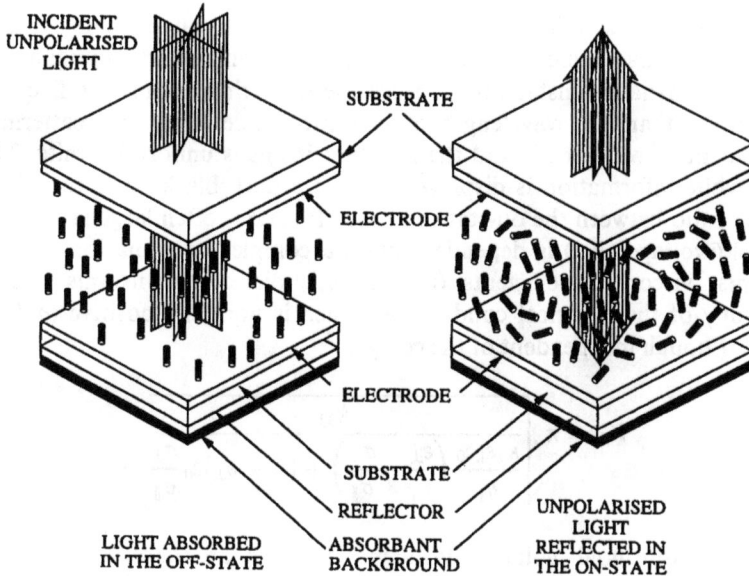

Figure 3.2 *Schematic representation of a dynamic scattering mode (DSM) LCD.*[1]

3.2. The substrates are coated with transparent electrodes (nesa or ITO) and often, but not always, with an appropriate alignment layer, usually a surfactant such as lecithin. A specularly reflecting mirror with a black background is fixed to the second substrate. No polarisers are required. Therefore, DSM-LCDs are intrinsically bright display devices with high contrast.

Off-State

Incident light striking a non-addressed part of a DSM-LCD with homeotropic alignment is transmitted unchanged by the homeotropically aligned nematic mixture, which is essentially transparent, and is then absorbed by the black background. Therefore, the cell appears dark in the non-addressed state (off-state). If the incident light is said to propagate along the z-axis then the refractive indices of the homeotropically aligned nematic medium in the x- and y-axes are equal. Therefore, the ordinary and the extraordinary rays of light are slowed down equally and no directional optical effect is apparent.

On-State

The application of a voltage with a magnitude above the threshold voltage, *e.g.* $5 V \mu m^{-1}$, across an activated pixel results in the scattering of incident light at that pixel. This is due to a circular stirring movement caused by the movement of positive and negative ions in opposite directions within the liquid crystal layer upon the action of the applied electric field. Momentum is transferred by friction to the liquid crystal, which results in turbulence and alignment discontinuities. The alignment discontinuities which cause changes in the effective refractive indices of adjacent nematic domains are designated as electrohydrodynamic instabilities.[2] Since the scattering centres are five to 10 times larger than the wavelength of incident visible light, the scattering is independent of wavelength and the resultant image is uniformly white. Thus, bright white information is displayed against a dark black background. The contrast ratio between the on-state and the off-state is often large (100:1). The threshold voltage (V_{th}) is dependent upon a complex combination of elastic constants, viscosity coefficients (κ_1 and η_1), dielectric constants and the anisotropy of the electrical conduction, which is normally positive ($\sigma_\parallel/\sigma_\perp \approx$ 4/3),[3,4] although independent of the cell gap.[3]

$$V_{th} = \pi \sqrt{\dfrac{k_{33}}{\dfrac{\kappa_1 \varepsilon_\parallel \varepsilon_0}{\eta_1}\left(\dfrac{\varepsilon_\perp}{\varepsilon_\parallel} - \dfrac{\sigma_\perp}{\sigma_\parallel}\right) + (\varepsilon_\parallel - \varepsilon_\perp)\varepsilon_0 \dfrac{\sigma_\perp}{\sigma_\parallel}}} \tag{1}$$

Below this threshold voltage, Williams domains are observed as visible birefringent stripes. As the voltage is increased further, more light is scattered and consequently the pixel appears brighter. This gives rise to grey-scale and the possibility of full colour, although typical operating voltages are relatively high (20–30 V). The switch on time, t_{on}, is inversely proportional to the current (I) and is generally short ($\approx$ 1–5 ms), even at high operating voltages (50–100 V $\approx$ 5–10 V μm^{-1}).

$$t_{on} \propto I^{-1} \tag{2}$$

On removal of the electric field, the ionic movement decreases dramatically and the initial alignment is recreated by elastic forces propagated from the surface boundary layers as well as by conduction. Therefore, switch-off times, t_{off}, are an order of magnitude longer ($\approx$ 20–100 ms) due to the high degree of disorder caused by the flow of liquid crystal and the absence of a restoring field effect.

$$t_{off} \propto \dfrac{d^2}{\sqrt{\sigma}} \tag{3}$$

The life-times of such displays depend on the type of applied potential: for AC operation, values of in excess of 10 000 operating hours have been reported, whereas DC potentials often lead to deposition of ions, with a consequent decrease, often dramatic, in life-time. The stability of the nematic material to electrochemical effects is also of critical importance.

Nematic Materials

The DSM effect was found[1] at the RCA Laboratories using individual Schiff's bases, see Table 3.1, which had been originally synthesised at the beginning of

Table 3.1 *Transition temperatures (°C), dielectric anisotropy (Δε) and conductivity ($\Omega^{-1} cm^{-1}$) for nematic liquid crystals (1–11) and a commercial mixture (Merck) designed for DSM-LCDs*

Molecular structure	Cr	N	I	Δε	σ	Ref
1 CH₃O–⟨⟩–CH=N–⟨⟩–O–CO–CH₃	• 83 •	100 •		−3.5	5 × 10⁻⁹	8
2 C₄H₉–⟨⟩–CH=C–⟨⟩–OCH₃	• 116 •	121 •				10
3 C₄H₉–⟨⟩–C=N–⟨⟩–OCH₃	• 108 (•	70) •				10
4 C₄H₉–⟨⟩–C≡C–⟨⟩–OCH₃	• 49 (•	37) •				11
5 C₄H₉–⟨⟩–CO–O–⟨⟩–OCH₃	• 41 (•	25) •		−0.4		12
6 C₄H₉–⟨⟩–N=N–⟨⟩–OCH₃	• 32 •	47 •		+0.2		13,14
7 C₄H₉–⟨⟩–N=C–⟨⟩–OCH₃	• 20 •	47 •				15
8 C₇H₁₅O–⟨⟩–N=N→O–⟨⟩–OC₇H₁₅	• 74 •	124* •		0.0	10⁻¹⁰	16
9 C₄H₉–⟨⟩–N=N→O–⟨⟩–C₄H₉	• 14 •	28 •		+0.2		17
10 C₄H₉–⟨⟩–N=N→O–⟨⟩–OCH₃ 6	• 16 •	76 •		+0.2	10⁻⁹	18
11 CH₃O–⟨⟩–N=N→O–⟨⟩–C₄H₉ 4						

Parentheses represent a monotropic transition temperature.
*SmC-N transition at 95 °C.

the 20th century.[8] The first report of the DSM-LCD described dynamic scattering of incident light at elevated temperatures within the nematic phase range of *N*-(4-methoxybenzylidine)-4′-acetoxyaniline (**1**),[8,9] see Table 3.1, above the melting point (83 °C), but below the clearing point (100 °C). However, this publication soon stimulated attempts to synthesise new nematic materials with a high clearing point, but also with a much lower melting point in order to enable the fabrication of commercial DSM-LCDs functional over a wide temperature range, including room temperature. The molecular structure of these new chemical entities exhibits a remarkable similarity which is characterised by two 1,4-disubstituted phenyl rings joined by an unsaturated linking unit and with an alkyl or alkoxy chain at each end of the molecule. Typical examples[8–17] (1–11) reported in 1971 or 1972 are collated in Table 3.1. The combination of a butyl and a methoxy terminal substituent were found generally to give rise to the lowest melting point of a wide series of homologues, see also Table 2.1 in Chapter 2. The nematic phase of these materials is generally of very slightly negative or positive dielectric anisotropy. However, most of these classes of compounds possess at least one seriously disadvantageous property for electro-optical applications: azobenzenes and azoxybenzenes are coloured and undergo *trans–cis* isomerisation; Schiff's bases are susceptible to decomposition by atmospheric moisture; *trans*-stilbenes undergo photocatalysed isomerisation to non liquid-crystalline *cis*-stilbenes which are subject to oxidation by atmospheric oxygen; phenyl benzoates are susceptible to photochemical instability due to photo-Fries reactions.

The material *N*-(4-methoxybenzylidine)-4'-butylaniline (MBBA; 7)[15] was the first compound synthesised with an enantiotropic nematic phase at room temperature. Compounds such as MBBA and 4,4'-dibutylazoxybenzene (9),[16,17] which also exhibits a nematic phase at room temperature, greatly facilitated the development of DSM-LCD prototypes as well as a better understanding of the mechanism of the DSM effect. However, the nematic clearing point of both compounds is too low for commercial applications. Therefore, nematic mixtures consisting of several components with a low melting point and a high clearing point, *e.g.* Licristal N4, Merck, a (6:4) binary mixture of 4-butyl-4'-methoxyazoxybenzene (10) and 4-methoxy-4'-butylaz-oxybenzene (11),[18] were then made commercially available, see Table 3.1. These mixtures exhibited a wide temperature range, high ionic conductivity, a positive electrical conduction anisotropy and weakly negative dielectric aniso-tropy. However, the first commercial DSM-LCDs commonly used mixtures containing ionic impurities, resulting from insufficient purification procedures or chemical decomposition. This often resulted in an inconsistent electro-optical performance, electrochemical decomposition of the nematic mixture itself and finally dielectric breakdown in the cell.[19] More reliable displays with longer lifetimes were produced by the use of improved commercial nematic mixtures with lower intrinsic conductivity ($5 \times 10^{-10} \, \Omega^{-1} \, cm^{-1}$) doped with an appro-priate amount of ionic material in order to produce the higher conductivity required for satisfactory DSM-LCD performance. The unwanted injection of additional ions from the electrodes could then be minimised by the use of AC instead of DC voltages.

Commercial DSM-LCDs generally utilised nematic liquid crystals of negative dielectric anisotropy. The exact role of the magnitude and sign of the dielectric anisotropy is somewhat unclear, although many investigations of this relation-ship were carried out.[20] Cells using nematic liquid crystals of positive dielectric anisotropy doped with ions and with a planar alignment give rise to a limited amount of dynamic scattering under certain conditions, but the major effect is dielectric realignment in electric fields.[21–23]

3 Cholesteric–Nematic Phase Change Effect (CNPC) LCDs[24–33]

The effect of magnetic and electric fields on chiral nematic structure was calculated[24,25] first and then confirmed by experiment for the magnetic case.[26,27] The confirmation of the effect of an electric field on the chiral nematic state led indirectly to the invention of a novel electro-optical display device based on the field induced cholesteric–nematic phase change effect, which was invented[28–30] at the Xerox Corporation Research Laboratories, Webster, New York, USA, at almost the same time as the DSM-LCD and GH-LCD, see Sections 2 and 7, at the RCA Corporation. Although based on light scattering in the same way as the DSM-LCD, the cholesteric–nematic phase change LCD is intrinsically less bright and exhibits a lower contrast ratio than corresponding DSM-LCDs due to the absorption of light of the first polariser. However, since the light scattering is caused by reflection from the helix of the chiral nematic phase as a kind of Bragg reflection and not by the electrohydrodynamic moment of ions, device lifetimes are intrinsically longer. There are several modifications of this basic operating principle.[31–34] This device is used most commonly in conjunction with dichroic dyes in order to improve the contrast, see Section 7.

Display Configuration

A chiral nematic (cholesteric) liquid crystal mixture is contained in a liquid crystal cell as shown in Figure 3.3.

Off-State

The helix of a chiral nematic liquid crystal of positive dielectric anisotropy is aligned parallel to the cell walls by application of a small electric field strong enough to overcome orientation effects due to anchoring at the cell boundaries. As the voltage is increased the pitch of the helix becomes larger, as more molecules are aligned parallel to the electric field. Eventually the helix is completely unwound and a pseudonematic phase of infinite pitch with the director parallel to the electric field vector is produced. Thus, application of an electric field across the cell has effected a phase change from the chiral nematic state to the nematic state. On removal of the field the optically active molecules in the chiral nematic mixture regenerate the helix in chiral nematic domains. However, there is no longer any macroscopic alignment of the helix axis in the cell and linearly polarised light passing through the first polariser is strongly scattered. Some of this scattered light is then transmitted through the second polariser and the display appears white in the off-state after the initial alignment treatment.

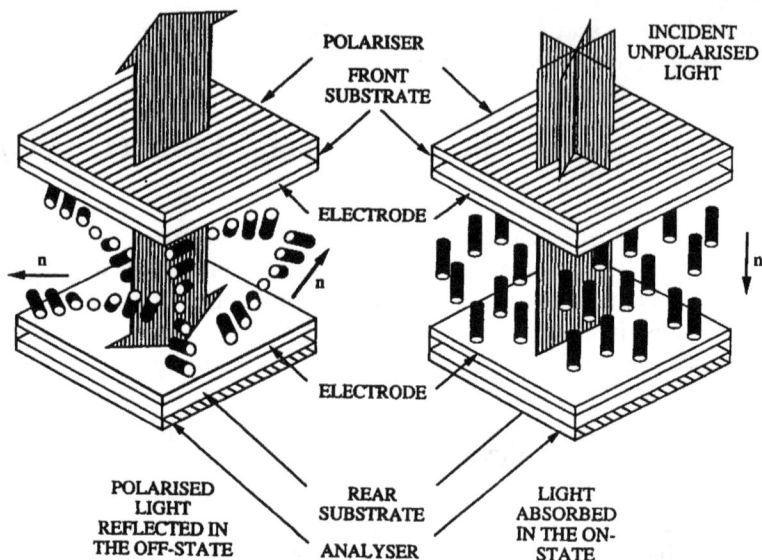

Figure 3.3 *Schematic representation of a cholesteric–nematic phase-change effect (CNPC) LCD.*[28-30]

On-State

When the helix is completely unwound a clear state is produced as a consequence of the formation of a pseudonematic phase with no helical structure from the corresponding chiral nematic phase. The analyser then absorbs linearly polarised light. Therefore, black information can be displayed at the appropriate pixels against a white background to produce an image with positive contrast. The threshold voltage for the unwinding of the helix is dependent on a number of physical parameters:

$$V_{th} = \frac{\pi^2 \sqrt{\dfrac{k_{22}}{\varepsilon_0 \Delta\varepsilon}}}{p_0 d} \tag{4}$$

where p_0 is the pitch length at zero field. It is found that relatively large electric fields (≈ 4–$10\,\mathrm{V}\,\mu\mathrm{m}^{-1}$) are required in order to completely unwind the helix. However, if a pitch is used that is longer than originally utilised, *e.g.* $3\,\mu\mathrm{m}$, much lower operating voltages can be achieved. Moreover, switching times are short, especially t_{off}, *e.g.* $60\,\mu\mathrm{s}$, since spontaneous formation of the chiral nematic helix takes place on removal of the electric field.[2]

Chiral Nematic Materials

The first prototype chiral nematic phase change devices were demonstrated with a tertiary mixture of cholesteryl chloride (**12**), nonanoate (**13**) and oleyl

Table 3.2 *Transition temperatures (°C) for chiral nematic derivatives of choles-terol (12–14) and a tertiary mixture (30%/56%/14%, respectively) for a CNPC-LCD prototype*[28–30]

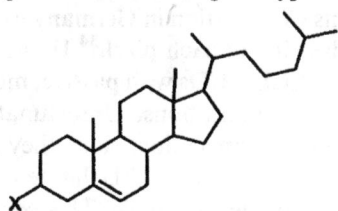

X		Cr		SmA*		N*		I
12	Cl	•	97		–	(•	62)	•
13	$C_8H_{17}CO_2$	•	80.5	(•	78)	•	92	•
14	$(Z)\text{-}C_8H_{17}C\!\!=\!\!CC_7H_{15}OCO_2$	•	50.5	(•	42)	(•	48)	•
Mixture		•	< 25		–	•	57	•

Parentheses represent a monotropic transition temperature.

carbonate (**14**). These three optically active derivatives of naturally occurring cholesterol, see Table 3.2, were of doubtful purity (98–99%), hydroscopic, of weak dielectric anisotropy and probably highly viscous, although no data were given. However, sample breakdown was reported above the clearing point, which is indicative of the presence of substantial amounts of chemical impurities.

Chemically, photochemically and electrochemically stable chiral nematic mixtures of low viscosity and strong positive dielectric anisotropy are required in order to produce CNPC-LCDs with fast switching times at low threshold and operating voltages and acceptable device lifetimes. This was soon achieved by doping nematic liquid crystals of positive dielectric anisotropy, see Section 5, with a suitable amount of optically active (chiral) dopant in order to induce the desired chiral nematic state with an appropriate pitch length. A chiral dopant with a very large helical twisting power allows small amounts to be used. This limits the increase in viscosity usually observed on addition of chiral dopants caused by chain branching at the optically active centre. It also keeps the change in the other physical properties of the host nematic mixture to a minimum.

4 Electrically Controlled Birefringence (DAP/HN/ECB) LCDs[34–39]

The many different kinds of LCDs[34–39] based on electrically controlled birefringence (ECB) are variants of the Fréedericksz effect first reported in the 1930s.[40] These related display types based on the same electro-optical effect, but a slightly different display configuration are denominated by a multitude of names. However, current versions of this type of display are now often referred to as deformation of vertically aligned phase DAP-LCDs or VAN-LCDs. They are nearly always combined with active matrix addressing.

The first LCDs based on electrically controlled birefringence reported by Schiekel and Fahrenschon of the Siemens Corporation in Germany were simple alpha-numeric displays with direct addressing of each pixel.[34] However, these early prototypes were soon superseded by ECB-LCDs with passive, multiplexed addressing for higher-information-content applications. Unfortunately, they were not particularly suited for this kind of application, since they generally exhibit low brightness ($\approx$ 20%), relatively low contrast (25:1), limited grey-scale and the observed colour varies strongly with viewing angle. This combination prohibited the large-scale commercialisation of ECB-LCDs with multiplex addressing. However, the advent of active matrix addressing moderated some of these disadvantages and ECB-LCDs with active matrix addressing are starting to appear on the market due to their capacity to generate full colour using grey-scale and relatively rapid response times compared to those of TN-LCDs with active matrix addressing. The development of stable nematic mixtures of negative dielectric anisotropy, high birefringence and low viscosity over recent years has made a significant contribution to the increasing atractivity of this type of LCD.

Display Configuration

A nematic liquid crystal of negative dielectric anisotropy is aligned with the director aligned orthogonal to the cell walls by means of a surfactant orientation layer, see Figure 3.4. One or two linear, elliptical or circular polarisers are

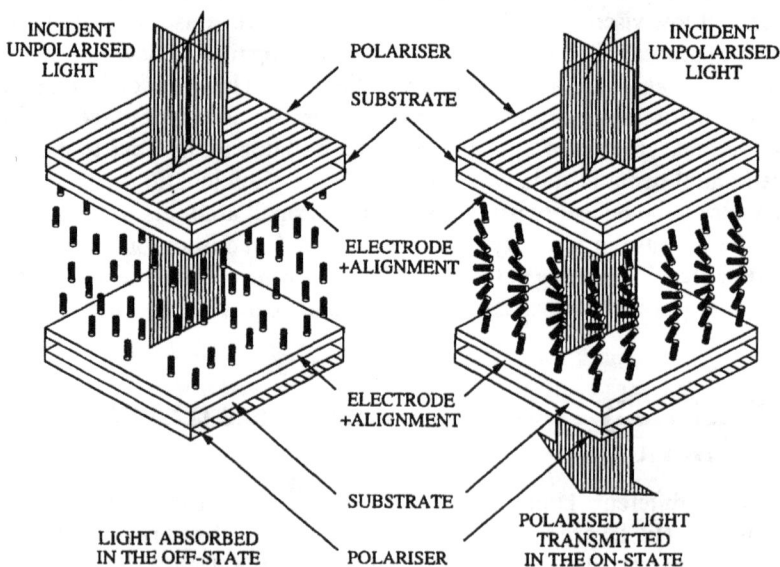

Figure 3.4   *Schematic representation of an electrically controlled birefringence (ECB) LCD.*[34-39]

required depending on the particular configuration of the device and whether the cell is operated in the transmissive or reflective mode.[39]

Off-State

If the cell is positioned between crossed polarisers then, when in the off-state, polarised light produced by the first polariser travels through the homeotropically aligned nematic medium unchanged, is absorbed by the analyser and the cell appears dark, see Figure 3.4. There is no effective birefringence ($\Delta n_{\text{eff}} = 0$) since $n_{\text{e}} = n_{\text{o}}$, see Chapter 2.

On-State

The application of a voltage above the threshold voltage, V_{th},

$$V_{\text{th}} = \sqrt{\frac{\pi^2 k_{33}}{(\varepsilon_\perp - \varepsilon_\parallel)\varepsilon_0}} \qquad (5)$$

where $\varepsilon_\perp - \varepsilon_\parallel < 0$, causes the nematic director to tilt away from the normal due to coupling of the electric field with the induced dielectric anisotropy.[41] This induces an effective birefringence, Δn_{eff}, which is proportional to the magnitude of the applied field up to the maximum value of $\Delta n = n_{\text{e}} - n_{\text{o}}$, see Chapter 1. Therefore, a proportion of the polarised light traverses the cell in the on-state. The proportion increases with increasing magnitude of the induced effective birefringence. However, since the optical effect is produced by interference between the extraordinary and the ordinary rays, white incident light is usually converted into coloured light due to interference. The optical retardation, δ, is proportional to the induced effective birefringence:[42]

$$\delta = \frac{2\pi \Delta n_{\text{eff}} d}{\lambda} \qquad (6)$$

where, d is the cell gap, when operated in the transmission mode for a given wavelength of light λ. The refractive index of the ordinary ray (n_{o}) passing through the cell is not affected by the applied field. However, the magnitude of the refractive index of the extraordinary ray (n_{e}) increases with voltage due to the dielectric coupling of the nematic director with the field, if the liquid crystal mixture is of negative dielectric anisotropy. Thus, the effective birefringence increases with applied electric field strength. In theory, this should allow the production of coloured displays by manipulation of the degree of optical retardation. However, in practise there is virtually no change in colour at voltages near the threshold voltages. Therefore, high-information-content ECB-LCDs with full colour are usually operated at voltages near the threshold voltage in combination with standard RGB colour filters.

The intensity, I, of light traversing the cell also depends on the optical retardation (phase difference), δ, as well as the angle of incidence of light (ϕ):

$$I \propto I_0 \sin^2(2\psi_0) \sin^2\left(\frac{\delta}{2}\right) \tag{7}$$

where I_0 is the intensity of the incident plane polarised light. The intensity of transmitted light is a maximum when

$$\Delta nd = (2m+1)\frac{\lambda}{2} \tag{8}$$

where m is an integer. Usually, Equation 8 is satisfied in the operating voltage region with $m = 0$ and $\lambda \approx 550\,\text{nm}$, which corresponds to the central wavelength of the visible spectrum and the wavelength of maximum sensitivity of the human eye. These conditions result in a black-and-white display with good uniform transmission across the visible spectrum. Since the induced birefringence depends on the strength of the electric field, the amount of light transmitted can be modulated in order to produce tuneable birefringence, *i.e.* grey-scale. Since the magnitude of the displacement of the director is not large for an optical effect to be observed, the switch-on times of ECB-LCDs are very short ($t_{on} \approx 20\,\mu\text{s}$). However, such short switch-on times are exhibited at relatively high operating voltages. The switch-off time, t_{off}, is a more complex parameter and depends on a viscosity coefficient, η, the cell gap, d and a curvature elastic modulus, κ:[41]

$$t_{off} \propto \frac{\eta d^2}{\kappa} \tag{9}$$

Therefore, switch-off times are independent of the field strength and directly dependent on material parameters, such as viscosity coefficients and elastic constants, and the cell configuration. Therefore, they are often three or four orders of magnitude larger than the switch-on times. However, sophisticated addressing techniques can produce much shorter combined response times ($t_{on} + t_{off} \approx 50\,\mu\text{s}$).[43,44] The nematic director should be inclined, *e.g.* 1° pretilt, in one specific direction in the zenithal axis in order to produce a uniform optical appearance which, however, may reduce the sharpness of the threshold voltage.[45]

Nematic Materials of Negative Dielectric Anisotropy

High-information-content ECB-LCDs require a sharp threshold voltage, low operating voltages and a steep electro-optical contrast curve.[46] Therefore, a high negative value of $\Delta\varepsilon$ and high positive values of k_{33}/k_{11}, Δn and δ are required. However, high values of Δn, which allow thin cells and, therefore, shorter response times, see Equation 9, also give rise to a residual birefringence in the off-state, due to the finite pretilt, and interference colours. This is responsible for the low contrast observed for thin ECB-LCDs. Unfortunately, a large cell gap generates long response times according to Equation 9. $\Delta\varepsilon/\varepsilon_\parallel$ should also be low, *i.e.* both $\varepsilon_\perp$ and $\varepsilon_\parallel$ should be high. However if $\Delta\varepsilon/\varepsilon_\parallel$ is too low, high operating voltages, which are already relatively high for operation with standard, low-cost CMOS drivers, are the result.

The first ECB-LCDs used nematic Schiff's bases of negative dielectric anisotropy, such as *N*-(4-methoxybenzylidene)-4'-butylaniline (**7**), see Table 3.1, or mixtures of homologues of this class of compound, which generally exhibit a nematic phase, a high clearing point and a melting point at or below room temperature. However, the dielectric anisotropy of these Schiff's bases is low ($0 > \Delta\varepsilon > -1$) due to the absence of any strong dipole moments orthogonal to the molecular axis. Consequently, the operating voltages of prototypes ECB-LCDs[34] were relatively high (≈ 7–8 V). The dependence of the threshold voltage on the elastic constants of these compounds could not be determined, since they were not known at the time.

Subsequent LCD types using electrically controlled birefringence, such as homeotropic nematic (HN-LCD)[39] or electrically controlled birefringence (ECB-LCDs),[36,38] used liquid crystal mixtures with a greater magnitude of the dielectric anisotropy with a negative sign. These mixtures incorporated components, such as compound (**15**) shown in Table 3.3, possessing a cyano group (CN) in a lateral position.[47] In this way the vector component of the dipole moment (4 D) of cyano group perpendicular to the long molecular axis was greater than that parallel to the long molecular axis.[47–55] However, substantial steric effects due to the large van der Waals volume of the cyano group induces substantial clearing point depressions as well as a high viscosity, while the induced negative dielectric anisotropy is only moderate ($-2 > \Delta\varepsilon > -4$).[49] This value is too low for practical ECB-LCDs, since it results in high operating voltages (see Equation 5).

Materials incorporating two cyano groups in adjacent lateral positions,[52,55] *e.g.* compound **16**,[48] exhibit a much greater negative value for the dielectric anisotropy ($-12 > \Delta\varepsilon > -20$); see Table 3.3. This is due to the large resultant dipole moment (≈ 7 D) orthogonal to the molecular long axis of the molecule, since the dipole vectors parallel to the long molecular axis cancel each other out. The viscosity does not increase proportionately for dicyano-substituted derivatives compared to that of corresponding mono-cyano-substituted derivative, although the resulting viscosity is very high. The clearing point may be even higher than that of the corresponding mono-cyano-substituted material. This may well be due to the shielding effect of the first nitrile group, since the molecular rotation volume of the mono- and dicyano-substituted compounds are almost identical. Hence, the degree of molecular separation due to steric effects of the lateral-substituent is very similar.

In the case of two-ring compounds the steric effects due to the presence of large lateral substituents induce a very low clearing point and a very high viscosity. The corresponding three-ring materials are also viscous. However, the presence of another ring often leads to observable mesomorphism in three-ring

Table 3.3 *Molecular structures, transition temperatures (°C) and dielectric anisotropy (Δε measured at 20°C) for nematic liquid crystals (15–27)*

Molecular structure	Cr	Sm	N	I	Δε	Ref
15	• 59	–	• 113	•	−3.8	47
16	• 138	–	• 148	•	−19.0	48
17	• 106	–	(• 102)	•	−11.5	52–55
18	• 88	• 154[a]	• 169	•		50
19	• 230	–	–	•		50
20	• 224	–	–	•		56
21	• 63	(• 43)	• 79.5 •			57
22	• 71.5 •	162	• 173 •			58
23	• 59.5	–	(• 0)	•	−6.0	59–61
24	• 25	• 30[b]	• 66	•	−8.0	62, 63
25	• 74	• 86[b]	• 171	•	−4.1	64
26	• 112	(• 105)[b]	• 190	•	−5.3	65
27	• 59	–	• 135	•	−5.1	66

[a] SmA. [b] SmB.
Parentheses represent a monotropic transition temperature.

compounds incorporating the nitrile function in a lateral position, *e.g.* compound **15**,[47] due to a higher degree of anisotropy in the molecular length/ breadth ratio and polarisability. Nematic mixtures of high negative dielectric anisotropy for LCD applications were made available in the late 1970s by the Chisso Corporation of Japan. Liquid crystalline ester derivatives of 4-*n*-alkoxy-2,3-dicyanophenol, such as compound **16**, were used to induce a negative value of the dielectric anisotropy.[48] Unfortunately, these mixtures exhibited an unacceptable degree of photochemical instability, which severely limited the lifetime of ECB-LCDs incorporating them. The high viscosity of these mixtures also gave rise to long response times, see Equation 9. The replacement of the terminal oxygen atom of 4-*n*-alkoxy-2,3-dicyanophenol by a methylene (CH_2) group to produce analogous 4-*n*-alkyl-2,3-dicyanophenyl ester derivatives, such as compound **17**, generated nematic materials of strong negative dielectric anisotropy which were sufficiently stable for commercial applications.[52,53] However, mixtures incorporating these substances were still very viscous with long response times in ECB-LCDs. Attempts to produce lower viscosity compounds by incorporating two cyclohexane rings, *e.g.* compound **18**,[50] or replacing the carboxy function by an ether link, *e.g.* compound **19**,[50] or by an ethyl group, *e.g.* compound **20**,[56] created almost completely insoluble, non-mesomorphic compounds with very high melting points. The incorporation of only one cyano group in non-ester, three-ring compounds via sp^2 bonding on aromatic rings (**21**)[57] and sp^3 bond angles in central linkages or terminal chains (**22**)[58] generated materials with nematic and smectic phases, but unfortunately with low negative values of dielectric anisotropy.

A diverse range of liquid crystalline 3,6-disubstituted pyridazines, such as compound **23**, whose lone-pair electrons situated at the electronegative nitrogen atoms exhibit a large resultant dipole moment perpendicular to the molecular long axis, see Figure 3.5, have also been prepared.[59-61] The lone-pair electrons do not increase the breadth of the molecule. Therefore, due to the absence of steric effects associated with lateral substituents, two-ring 3,6-disubstituted pyridazine derivatives exhibit nematic phases of high negative dielectric aniso-tropy ($-6 > \Delta\varepsilon > -10$) and much lower viscosity than dicyano-substituted compounds with a dielectric anisotropy of similar magnitude. Unfortunately, they also suffer from the serious disadvantages for commercial ECB-LCDs of photochemical instability and significant dark-current conductivity.

The *trans,trans*-4,4'-dialkyl-1,1'-dicyclohexyl-4'-carbonitriles, *e.g.* compound **24**, incorporate only one cyano group like the mono-nitriles (**15**, **21** and **22**). However, the sp^3 bonding and the axial position of this nitrile function on one of the carbon atoms of one of the cyclohexane rings results in the dipole moment being almost exactly perpendicular to the molecular long axis, see Figure 3.6.[62,63] The axial hydrogen atoms on the same side of the cyclohexane ring shield the axial cyano group to some degree. Consequently, these chemically, thermally and photochemically stable materials exhibit an enantiotropic nematic phase over a wide temperature range. The viscosity of the bicyclo-hexyl-nitriles is much lower than that of the aromatic, monocyano-substituted

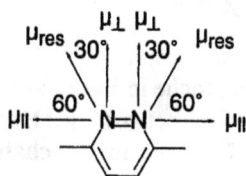

Figure 3.5   *Schematic representation of the resultant transverse dipole moment across the molecular long axis of a 3,6-disubstituted pyridazine derivative.*

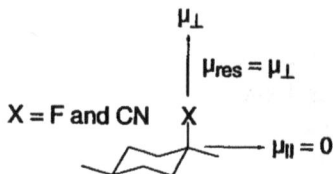

Figure 3.6 *Schematic representation of the resultant transverse dipole moment across the molecular long axis of an axially 4'-fluoro-substituted or 4'-cyano-substituted trans,trans–4,4'-dialkyl-1,1'-dicyclohexane derivative.*

compounds. Unfortunately, the very low birefringence of axially substituted bicyclohexanes renders them unsuitable as ECB-LCDs. However, related derivatives incorporating two benzene rings and one axially cyano-substituted cyclohexane ring exhibit a birefringence of sufficient magnitude compatible with ECB-LCD applications.

2,3-Difluorobenzene derivatives such as compound **25**,[64] see Table 3.3, embody the best combination of physical properties of nematogens of negative dielectric anisotropy and low viscosity for ECB-LCDs. The small van der Waals volume of the fluorine atom (5.8 Å) and its relatively small dipole moment (1.47 D) lead to materials of relatively low viscosity and weak to moderately strong negative dielectric anisotropy in the nematic phase. In diether derivatives of 2,3-difluoroquinone the electronegative fluorine atoms polarise the additional π-electron density provided by the conjugated lone-pair electrons on the oxygen atoms. The large resultant dipole moment is perpendicular to the molecular long axis. Thus, relatively large values of the dielectric anisotropy ($-4 > \Delta\varepsilon > -6$) are observed. A similar effect is observed for the fluoro-substituted pyridines, such as compound **26**, which also exhibit a strongly negative dielectric anisotropy due to the resultant dipole moment orthogonal to the long molecular axis.[65] However, these synthetically more complex compounds appear to exhibit comparable properties to those of the analogous 2,3-difluorophenyl derivatives. Some attempts to improve the properties of 2,3-difluorophenyl derivatives by synthesising compounds with different central linkages, *e.g.* compound **27**, were relatively successful.[66] A range of diether derivatives of 2,3-difluoroquinone are manufactured commercially (E. Merck) for a several types of LCDs, which require nematic mixtures of negative dielectric anisotropy, especially ECB-LCDs.

5 Twisted Nematic LCDs[67]

The twisted nematic liquid crystal display (TN-LCD) was reported by Schadt and Helfrich of F. Hoffman-La Roche in Basle, Switzerland in 1970.[67] This was part of a tripartite collaboration between F. Hoffman-La Roche in Basle, Brown Boveri of Baden and Ebauche in Neuchatel, all in Switzerland. The intention was to design and develop flat panel displays, *e.g.* for digital watches. The first LCD factory was constructed in Lenzburg, Switzerland in the mid-1970s by Videlec, a subsidiary of Brown Boveri. Since then the TN-LCD has established itself as the dominant electro-optical display by far in portable instruments and is slowly gaining market share in the computer monitor market. TN-LCDs with direct or multiplexed addressing are to be found in low-information-content electro-optical devices such as digital clocks, watches, computer games and calculators. TN-LCDs with active-matrix addressing are steadily displacing multiplexed STN-LCDs, see Section 6, from the high-information-content displays market, *e.g.* in notebooks, personal organisers or laptop computers. The dominance of the TN-LCD with different modes of addressing is due to an advantageous combination of properties for a given application selected from low power consumption, low threshold and operating voltages, acceptable contrast (at least for narrow viewing angles), relatively efficient multiplexing (at·least at low ratios, *e.g.* a 1:32 duty cycle), a long operating lifetime and exceptionally low cost. The main drawbacks of the TN-LCD are long response times, especially at low temperatures, poor brightness, limited contrast and a large variation of contrast with viewing angle, especially in applications with even a moderate degree of multiplexed addressing.

Display Configuration

A standard TN-LCD consists of a nematic liquid crystal mixture of positive dielectric anisotropy contained in a cell with an alignment layer on both substrate surfaces, usually rubbed polyimide, crossed polarisers and a cell gap of 5–10 μm, see Figure 3.7. The nematic director is aligned parallel to the direction of rubbing in the azimuthal plane of the device. The alignment layer induces a small pretilt angle ($\theta \approx 1$–3°) of the director in the zenithal plane. The

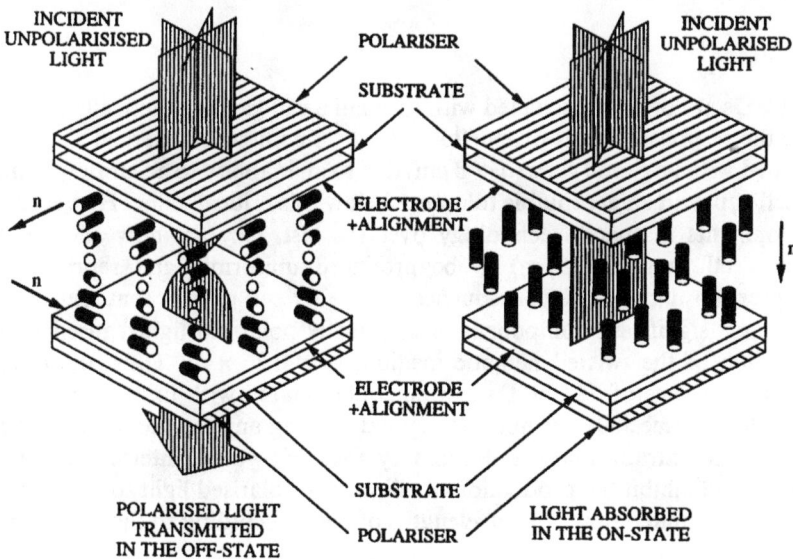

Figure 3.7 *Schematic representation of a twisted nematic liquid crystal display*[67] *(TN-LCD).*[169]

direction of alignment at the upper glass substrate is orthogonal to that at the lower glass substrate, *i.e.* there is an overall twist of 90° in the director from the top to the bottom of the display maintained by elastic forces in the off-state. The polarisation axes of the polarisers are usually arranged parallel to the director of the nematic medium at each substrate surface. Occasionally the polarisers are fixed parallel to one another in order to generate a dark off-state and a bright on-state with a wide viewing angle cone.

Off-State

The plane of polarised light generated by passage through the first polariser is rotated by 90° by the twisted nematic medium due to a wave-guiding action as long as the wavelength of light (380–780 nm) transmitted is much less than the propagation depth ($d \approx 5$–$10\,\mu$m) of the nematic layer.[68] This waveguiding action allows the plane polarised light to pass through the second polariser (analyser), since the plane of linear polarised light and the transmission axis of the analyser are now parallel. Consequently, the off-state, or transition-state, is white and is often referred to as the white- or open-mode. In the reflective TN-LCD configuration, transmitted, wave-guided light is reflected back along the same pathway by a reflective surface.

In order to suppress the formation of elliptically polarised light rather than plane polarised light in a TN-LCD, it is necessary to control the cell thickness, d, and the birefringence of the nematic medium, Δn, with respect to the wavelength of light in a vacuum, λ, according to the equation:

$$d\Delta n \gg \frac{\lambda}{2} \qquad (10)$$

TN-LCDs are often constructed with as small a cell gap as practically possible, since the response time is proportional to d^2. The first commercial TN-LCDs operated with large cell gaps ($d \approx 8\,\mu$m) due to an inability to accurately control the cell gap with an acceptable tolerance below this critical value. However, the developments in spacer technology over the last 25 years now allow much smaller cell gaps ($d \approx 5\,\mu$m) to be produced uniformly and reproducibly. However, an undesirable consequence of smaller cell gaps is that transmitted light has a significant component of elliptically polarised light caused by the tight pitch of the twisted nematic medium ($p = 0.25 \times d$). Thus, the desired white appearance of TN-LCDs can be contaminated with undesired interference colours, especially for non-orthogonal viewing angles, due to interference between the extraordinary and ordinary rays. However, interference at the analyser can inhibit the production of elliptically polarised light to a significant degree for a limited range of wavelengths of light, according to the equation

$$u = \frac{2d\Delta n}{\lambda} \qquad (11)$$

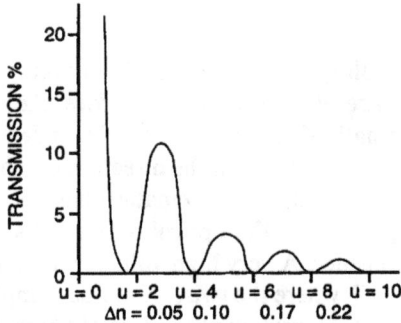

Figure 3.8 *Gooch and Tarry plot of transmission* versus *the coefficient, u, and the corresponding birefringence values, Δn, for a Twisted Nematic Liquid Crystal Display TN-LCD of thickness 6.55 μm operating between parallel polarisers.[69,169]*

where the coefficient u must be larger than unity, which is the so-called Maugin limit.[68] Thus, for the centre of the spectral range of visible light ($\lambda \approx 550$ nm), the wavelength at which the eye is most sensitive, a series of maxima and minima in transmission are produced for certain values of u (*i.e.* $\sqrt{3}$, $\sqrt{15}$, $\sqrt{35}$, $\sqrt{63}$, *etc.*), see Figure 3.8. These correspond to the first, second, *etc.*, Maugin, or Gooch and Tarry, minima.[68,69] Therefore, for a TN-LCD assuming ideal parallel polarisers and neglecting light absorbed by the first polariser, the transmission:

$$I \propto \sin^2\left[\frac{\pi/2(1+u^2)^{1/2}}{1+u^2}\right] \qquad (12)$$

where the amplitude of transmission decreases with increasing value of u. Therefore, TN-LCDs should be operated in the first or second minimum in order to optimise brightness, contrast and the dependence of the contrast on viewing angle. The minima become maxima and *vice versa* for crossed polarisers. A TN-LCD operated in the first minimum with a moderately small cell gap ($d = 6.55$ μm) and a nematic liquid crystal mixture of a low birefringence ($\Delta n < 0.10$) exhibits a bright appearance, see Figure 3.8.[70] A nematic mixture of high positive dielectric anisotropy and low viscosity is used to achieve fast response times. Higher values of the birefringence at constant cell gap correspond to higher Gooch and Tarry minima with a consequent lower transmission in the off-state. A certain amount of light still leaks through all TN-LCDs in the off-state due to non-ideal polarisers, non-uniform nematic alignment, cell thickness variation, temperature dependence and dispersity of Δn and the polychromaticity of light. A combination of all of these factors results in a lower contrast between the on- and off-state than calculations would suggest.

On-State

Upon application of a voltage above the threshold voltage, V_{th}, in a TN-LCD the molecules in the centre of the cell and, therefore, the optical axis, which is coincident with the nematic director, start to tilt under the influence of the electric field. They realign themselves in the direction of maximum polarisation, *i.e.* parallel to the electric field, if the nematic phase is of positive dielectric anisotropy. This realignment of the optical axis results in a decrease in the effective average birefringence. A very high voltage would lead to a tilt angle of 90° at the centre of the cell, where the twist would be completely unwound. The effective birefringence is still very low at moderately high applied voltages and the plane of the polarised light is no longer rotated. The plane polarised light strikes the analyser in the crossed polariser configuration and is absorbed. A TN-LCD in this configuration exhibits positive contrast, *i.e.* displays black information on a white background. The threshold voltage for a TN-LCD is defined as:[67,71]

$$V_{th} = \pi \left[\frac{1}{\varepsilon_0 \Delta \varepsilon} \left(k_{11} + \frac{k_{33} - 2k_{22}}{4} \right) \right]^{1/2} \qquad (13)$$

where k_{11}, k_{22} and k_{33} are the Frank elastic constants, ε_0 is the dielectric constant of a vacuum and $\Delta \varepsilon$ is the dielectric anisotropy of the nematic medium. The switch-on time, t_{on} is defined by:[67]

$$t_{on} \propto \frac{\eta d^2}{\Delta \varepsilon E - \kappa \pi^2} \qquad (14)$$

In the absence of a distorting applied field, surface and elastic forces regenerate the original twist in the nematic director and plane polarised light is once again transmitted. Several layers of the nematic medium several molecules thick do not realign at all under the action of an electric field due to strong anchoring with the orientation layer. Although these residual layers of bend and splay at the cell surfaces lower the contrast, they transmit the anchoring of the director at the substrate surfaces and are responsible for the regeneration of the twisted nematic configuration in the absence of an applied electric field. Therefore, the switch-off time, t_{off}, is defined as:[67]

$$t_{off} \propto \frac{\eta d^2}{\kappa \pi^2} \qquad (15)$$

where

$$\kappa = \left[k_{11} + \left(\frac{k_{33} - 2k_{22}}{4} \right) \right] \qquad (16)$$

and η is the rotational viscosity, γ_1, and E is the applied field. The response of the director to the applied electric field is clearly not identical throughout the cell and depends on the strength of the anchoring at the substrate surfaces. Thus, the director at the centre of the cell, *i.e.* furthest from the substrate surfaces, responds first.

The contrast of a TN-LCD is highest parallel to the optical axis in the middle of the cell and lowest at angles normal to this direction. Therefore, the contrast of a TN-LCD is not symmetrical and depends not only on the viewing angle, but also on the direction of view with respect to the director, see Chapter 2. In the absence of a chiral dopant, there are also two degenerate, but optically non-equivalent, directions of 90° twist of the director in the off-state ($+90°$ and $-270°$). This gives rise to domains of opposite (reverse) twist upon removal of the applied electric field. Disclination lines between adjacent domains of opposite twist reduce the contrast. This can be eliminated by the addition of a small amount of an optically active compound (chiral dopant),[72] which induces a uniform direction, or handedness of twist, in the nematic medium ($+90°$). However, if the ratio of the pitch and cell gap (p/d) is too small then areas of reverse twist can still be formed. Therefore, the pitch, p, induced in the nematic medium by the chiral dopant should be long corresponding to a relatively large ratio of the chiral nematic pitch to the cell gap ($p/d \approx 7$). An upper limit of the pitch is set by the fact that too large values for the pitch will fail to inhibit the formation of reverse-twist domains. An unintended consequence of the addition of a chiral dopant is that the medium is also no longer a twisted nematic phase, but a chiral twisted nematic phase, which exhibits a higher threshold voltage:

$$V_{th} = \sqrt{\frac{k_{11}\pi^2 + \left\{k_{33} - k_{22}\left(1 - \frac{4d}{p}\right)\right\}\left(\frac{\pi}{2}\right)^2}{\varepsilon_0 \Delta \varepsilon}} \qquad (17)$$

where the pitch of the chiral nematic mixture represents an additional term.[67] Initially derivatives of cholesterol, such as cholesteryl nonanoate (13), see Table 3.2[73] were added to induce the desired handedness of twist. However, ester derivatives of cholesterol are hygroscopic, relatively unstable and induce a high nematic viscosity at low concentration. Therefore, they were soon replaced by optically active versions of the components of the nematic mixture, *e.g.* 4-cyano-4'-[(S)-2-methylbutyl]biphenyl.[74,75]

Non-ideal birefringence effects are enhanced during grey-scale operation, *e.g.* for full-colour displays, since the optical axis of the nematic layer makes an oblique angle to the normal of the cell and the applied electric field. This results in a pronounced asymmetric viewing angle dependence of contrast, luminance (brightness) and chromaticity, due to interference effects, and even to inversion of contrast at large viewing angles. Contrast ratios above 10:1 are limited for a typical TN-LCD to a maximum viewing angle of $\pm45°$ in the horizontal plane and $+60°$ and $-15°$ in the vertical plane, see Chapter 2. This is clearly unacceptable for large area displays, *e.g.* for computer monitors and portable computers.

The large viewing angle dependency of TN-LCDs attributable to the positive birefringence of the nematic layer can be compensated to a large degree by using one or more optical retarders of positive or negative birefringence in combination with the normally white (NW) mode, *i.e.* with crossed polarisers, see Chapter 2. The NW mode of TN-LCDs is used in combination with active matrix addressing to generate near video-rate addressing for high-information-content, full-colour applications. The suppression of unwanted leakage of light in the on-state leads to higher contrast at wider viewing angles and suppresses grey-scale inversion. The asymmetric twist of the nematic medium is compensated for by aligning the optical axis of the two compensation layers on either side of the cell at right angles to each other, which results in no retardation at normal incidence. Therefore, optical compensation layers can significantly improve the contrast and the viewing-angle dependence of contrast in the vertical direction as well as in the horizontal direction for non normal viewing angles, while maintaining the good optical properties at normal viewing. However, these layers of organic material also absorb a proportion of the transmitted light and, therefore, they contribute to the intrinsically low brightness of LCDs. A higher initial intensity of incident light leads to a higher power consumption. This is undesirable for portable applications due to the shorter time between battery recharging times.

Nematic Materials of Positive Dielectric Anisotropy

Over the last 30 years since the invention of the TN-LCD[67] there has been a gradual, but steady increase in the size, complexity and information-content of TN-LCDs, represented in the number of pixels. A full-colour TN-LCD with active matrix addressing for a portable computer or computer monitor may contain three million pixels or more. The changes in the methods of electronic addressing, from direct addressing of individual pixels in watches and simple calculators, to multiplex addressing used in complex calculators, personal digital assistants and cellular telephones, to active matrix addressing essential for portable computers and computer monitors, resulted in a continual need to synthesise new classes of nematic liquid crystals to create nematic mixtures with the desired spectrum of physical properties. The major developments in the synthesis of nematic materials for TN-LCDs with the three different types of addressing are described below.

Nematic Materials for Direct Addressing

The operating principle of the TN-LCD was first demonstrated[67] at a temperature above 100°C between the melting and clearing point of *N*-(4-ethoxybenzylidene)-4-aminobenzonitrile (**28**), see Table 3.4. This Schiff's base was first prepared at the beginning of the 20th century in Vorländer's research group in Halle, Germany.[76] It was synthesised again later and was found to exhibit a positive dielectric anisotropy ($\Delta n \approx +14$).[77,78] The researchers at Roche then prepared a tertiary mixture of two 4-*n*-alkoxybenzylidene-4-aminobenzoni-

Table 3.4 *Transition temperatures (°C) and dielectric anisotropy for the aromatic nitriles (28–40)*

Molecular structure	Cr		N		I	$\Delta\varepsilon$	Ref
28 C_2H_5O— ... —CN	●	106	●	128	●	+14	77, 78
29 C_4H_9O— ... —CN	●	63	●	111	●		79
30 $C_6H_{13}O$— ... —CN	●	55	●	101	●		79
31 C_7H_{15}— ... —CN	●	54	●	98	●		80
32 C_4H_9— ... —CN	●	66	(●	41)	●		80
33 C_6H_{13}— ... —CN	●	46	●	55	●		80
34 C_3H_7— ... —CN	●	65.5	●	77.5	●		80
35 C_6H_{13}— ... —CN	●	33	●	64.5	●		80
36 C_5H_{11}— ... —CN	●	22.5	●	35	●	+8.5	82
37 C_5H_{11}— ... —CN	●	33.5	●	43.5	●		83
38 C_5H_{11}— ... —CN	●	73	[●	31]	●	+9.4	84
39 C_5H_{11}— ... —CN	●	71	(●	52)	●	+21	85
40 C_5H_{11}— ... —CN	●	96	●	109[a]	●	+3.4	85

Parentheses represent a monotropic transition temperature; brackets represent a virtual, extrapolated transition temperature; a SmA phase at 93.5°C.

triles[79] (**29** and **30**) and 4-heptanoylbenzylidene-4-aminobenzonitrile[80] (**31**). This simple nematic mixture of high birefringence and strong positive dielectric anisotropy enabled prototype TN-LCDs to be operated and evaluated at room temperature.[81] Mixtures of liquid crystalline compounds usually exhibit a much lower melting point than that of the individual components of the mixture, whereas the clearing point is usually an average, although some deviation from ideal behaviour is nearly always observed. Mixtures with the lowest melting point using the components chosen are referred to as eutectic mixtures. The composition of these mixtures can be predicted with a good degree of accuracy, if the melting point and the enthalpy of fusion of the individual components are known. The binary mixtures consisting of the esters (**32** and **33**) and the Schiff's bases (**34** and **35**) are also both nematic at room temperature.[81] However, the threshold voltage of both of these mixtures is much lower than that of the tertiary mixture of **29**, **30** and **31**. This is a direct consequence of a favourable combination of elastic constants (see Equation 13) as the dielectric anisotropy of all three mixtures is almost identical and the threshold voltage is independent of the cell gap. Furthermore, the shorter switching times observed for the binary mixture of the Schiff's bases (**34** and **35**) is also due to a lower rotational viscosity, γ_1, of these compounds compared to that of the esters **32** and **33** (see Equations 14 and 15) since the value of the elastic constant ratio κ of each of the binary mixtures is very similar. The viscosity of the binary mixture of alkyl-substituted Schiff's bases (**34** and **35**) is clearly lower than that of the tertiary mixture of the corresponding alkoxy-substituted or alkanoyloxy-substituted Schiff's bases (**29**, **30** and **31**).

The first commercially available TN-LCDs were manufactured in the mid-1970s using more sophisticated mixtures of Schiff's bases and phenyl benzoate esters supplied by F. Hoffmann-La Roche. The Schiff's bases had to be handled with some care due to the hydrolysis of the central linkage by atmospheric moisture. Researchers in the research group of G. W. Gray at the University of Hull in the United Kingdom realised in the early 1970s that the chemical, photochemical or electrochemical instability encountered for many liquid crystals known at that time (see Table 2.1) could be directly attributed to the conjugated, unsaturated central linkage between the aromatic rings. Therefore, they designed and synthesised the class of compounds known usually as the cyanobiphenyls, see Table 3.4 for a representative example of this important class of liquid crystalline compounds, *i.e.* 4-cyano-4'-pentylbiphenyl (**36**), which is nematic at room temperature. The 4-*n*-alkyl-4'-cyanobiphenyls do not contain a central linkage at all between the two phenyl rings.[82] It is not surprising, therefore, that they were found to be sufficiently chemically, photochemically and electrochemically stable for use in TN-LCDs with direct addressing. Furthermore, and perhaps critically, the absence of the central linkage induced a lower viscosity and melting point than those of the corresponding aromatic cyanophenyl benzoate esters and cyano-substituted Schiff's bases available at that time. As a consequence of this advantageous combination of physical properties, mixtures such as E7 and E8 (BDH Chemicals, now a subsidiary E. Merck, Darmstadt, Germany) of 4-*n*-alkyl-4'-cyanobiphenyls, 4-

n-alkoxy-4'-cyanobiphenyls and 4-*n*-alkyl-4"-cyano-*p*-terphenyls gradually displaced the mixtures of Schiff's bases and phenyl benzoate esters from most commercial TN-LCDs.

Nematic mixtures composed of these new classes of materials were used successfully in simple TN-LCDs, *e.g.* digital watches, clocks and calculators, where the information content is limited and each pixel can be driven directly with a dedicated electrode contact and individual driver, see Chapter 2. With this type of addressing the off-voltage is zero and the on-voltage may be several times larger than the threshold voltage. Therefore, in a TN-LCD, where the electro-optical characteristic is relatively flat, a good contrast can be attained as well as low power consumption. Nematic materials for this kind of application should, in the first instance, be chemically, electrochemically and photochemically stable. Furthermore, they should exhibit a low melting point, a high clearing point, a wide nematic phase, no smectic mesomorphism, especially no ordered smectic phases such as smectic *B* or *E*, a low viscosity, a high birefringence and a high positive value of the dielectric anisotropy. The compounds described below represent attempts to design and synthesise new classes of nematic compounds with an optimal combination of these physical properties for TN-LCDs.

The classes of nematic compounds represented in the Tables 3.4 and 3.5 by their pentyl homologues (**36–59**)[82–101] in order to facilitate comparison, are essentially modifications of the 4-*n*-alkyl-4'-cyanobiphenyls.[82] The dielectric anisotropy of the 2-(4-cyanophenyl)-5-pentylpyridine (**37**),[83] 2-cyano-5-(4-pentylphenyl)pyridine (**38**)[84] and the 2-(4-cyanophenyl)-5-pentylpyrimidine (**39**)[85] is higher than that of the 4-cyano-4'-pentylbiphenyl (**36**) without a nitrogen heterocycle. The higher value of the dielectric anisotropy of the heterocyclic compounds (**37–39**) is due to the resultant dipole moments associated with the lone pair electrons on the nitrogen atoms being directed along the long molecular axis, see Table 3.4. If the dipole moments are directed in the opposite direction, *e.g.* in the 5-cyano-2-(4-pentylphenyl)pyrimidines (**40**),[85] the magnitude of Δε is correspondingly lower. The 2-(4-cyanophenyl)-5-*n*-alkylpyrimidines[85] were synthesised on a commercial scale as components of nematic mixtures for TN-LCDs for many years by F. Hoffmann-La Roche.

The chemically, photochemically and electrochemically stable class of compounds referred to generally as the PCHs (pentylcyclohexanes) is exemplified by the *trans*-1-(4-cyanophenyl)-4-pentylcyclohexane (**41**), see Table 3.5.[87] This compound possesses a *trans*-1,4-disubstituted cyclohexane ring in place of the phenyl ring attached to the alkyl group of the structurally related 4-cyano-4'-pentylbiphenyl (**36**). However, PCH-5 (**41**) exhibits a lower birefringence, a lower viscosity and a comparable value of the dielectric anisotropy to those of the corresponding cyanobiphenyl (**36**). This particularly advantageous combination of physical properties inspired the synthesis of a wide range of related materials, *e.g.* compounds (**41–53**),[86–96] with one alicyclic ring bearing a terminal alkyl group and one aromatic ring bearing a cyano group. Several classes of the nitriles prepared, such as homologues of the dioxane (**49**)[93] have also been manufactured as components of nematic mixtures for TN-LCDs. In

Table 3.5 *Transition temperatures (°C) and some values for the dielectric anisotropy of the nitriles (41–59)*

	Molecular structure	Cr		N		I	Δε	Ref
41	C₅H₁₁—⬡—⬡—CN	●	31	●	55	●	+9.9ᵃ	86
42	C₅H₁₁—⬡—⬡(N)—CN	●	45.5	●	55.5	●		87
43	C₅H₁₁—⬡—⬡(N)—CN	●	48	●	64	●		88
44	C₅H₁₁—⬡—⬡(N,N)—CN	●	84	(●	36)	●		89
45	C₅H₁₁—⬡—⬡(N,N)—CN	●	70	●	98*	●		89
46	C₅H₁₁—N⬡N—⬡—CN	●	40		–	●		90
47	C₅H₁₁—⬡—⬡—CN	●	48	●	61	●		91
48	C₅H₁₁—⬡—⬡—CN	●	35	(●	5)	●		92
49	C₅H₁₁—⬡(O,O)—⬡—CN	●	56	(●	52)	●	+11ᵇ	93
50	C₅H₁₁—⬡(O,B,O)—⬡—CN	●	48		–	●		94
51	C₅H₁₁—⬡(O,S)—⬡—CN	●	74	(●	19)	●		95
52	C₅H₁₁—⬡(S,S)—⬡—CN	●	98		–	●		95
53	C₅H₁₁—⬡—⬡—CN	●	62	●	100	●	+10ᶜ	96
54	C₅H₁₁—⬡—⬡—CN	●	<25	●	−25	●		97
55	C₅H₁₁—⬡—⬡—CN	●	113	(●	50)	●		98
56	C₅H₁₁—⬡—⬡—CN	●	62	●	85†	●	+3~4	99

(continued)

Table 3.5 *continued.*

	Molecular structure	Cr		N		I	$\Delta\varepsilon$	Ref
57	C_5H_{11}—[ ]—N—[ ]—CN	●	57	●	65	●		100
58	C_5H_{11}—[ ]—[ ]—CN	●	46	●	47	●		101
59	C_5H_{11}—[ ]—[ ]—CN	●	104	●	129	●		98

Parentheses represents a monotropic transition temperature; * SmB phase at 94°C; $\Delta\varepsilon$ measured at: [a]$0.98 \times T_{NI}$; [b]42°C; [c]25°C.

the case of the dioxanes this is partly due to the large value of the dielectric anisotropy and low birefringence. The high $\Delta\varepsilon$ is due in part to the dipole moments associated with the electronegative oxygen atoms. The high dielectric anisotropy leads to lower threshold and operating voltages. The other classes of aromatic nitriles collated in Table 3.5 exhibit at least one unfavourable property, which inhibited their manufacture, such as the photochemical instability of conjugated cyclohexenes, *e.g.* compound **47**, low resistivity values due to ionic impurities of the borinane (**50**) or high viscosity, *e.g.* the bicyclo[2,2,2]octane (**53**).

If the cyano group is attached to a non-conjugated alicyclic ring, as in the aliphatic nitriles (**54–59**) also collated in Table 3.5,[97–101] then the dipole moment of the cyano group is much lower than of an analogous benzonitrile due to the absence of aromatic conjugation with the cyano group. Consequently, the value of the dielectric anisotropy is lower for aliphatic nitriles compared to that of the related conjugated benzonitriles also shown in Table 3.5. However, the *trans*-1-(*trans*-4-cyanocyclohexyl)-4-pentylcyclohexane (**56**) with two cyclohexane rings still possesses a moderately high value for the dielectric anisotropy in spite of the absence of polarisable aromatic π-electrons.[99] Since the nematic clearing point is high and the birefringence and viscosity values are low, homologues of the aliphatic nitrile (**56**), often referred to as CCH-5, are also used in nematic mixtures (E. Merck). They are used in particular as important components of nematic mixtures of low Δn for TN-LCDs with a moderate cell gap ($d \approx 6.55\,\mu m$) operated in the first minimum,[70] see Figure 3.8. The other aliphatic nitriles collated in Table 3.5 generally exhibit an unattractive combination of a high melting point, a low nematic clearing point and a high viscosity.

The *trans*-1-*n*-alkyl-4-(4-cyanophenyl)cyclohexanes[86] exhibited an unparalleled advantageous combination of physical properties for TN-LCDs at this time. Therefore, a central linkage was incorporated between the cyclohexane and the phenyl ring of PCH-5 (**41**) to form the structurally related compounds (**60–69**) collated in Table 3.6 in attempts to improve on these properties.[86,102–106] However, only the *trans*-1-*n*-alkyl-4-[2-(4-cyanophenyl)ethyl]cyclohexanes (PECHs), *e.g.* PECH-5 (**62**),[104] were almost equivalent from the point of view of

Table 3.6 *Transition temperatures (°C) for the benzonitriles (41 and 60–69)*

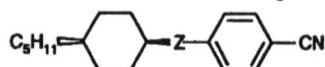

$$C_5H_{11} - \text{(cyclohexyl)} - Z - \text{(phenyl)} - CN$$

	Z	Cr		N		I		Ref
60		•	55	•	81	•		102
61		•	74	(•	67)[a]	•		103
41	—	•	30	•	59	•		86
62		•	45	•	55	•		104
63		•	74	(•	49)	•		104
64		•	54	•	107	•		105
65		•	41	•	73	•		105
66		•	51	(•	39)	•		106
67		•	40	(•	39)	•		106
68		•	63	(•	54)	•		106
69		•	64	(•	43)	•		106

Parentheses represent a monotropic transition temperature; [a]monotropic SmX–SmB transition at 58°C.

display devices. None of these materials are currently being manufactured, apart from the ester (**60**), which was prepared before the PCHs.[102]

Variations in the nature of the terminal chains was then also investigated, see Table 3.7, where a series of *trans*-4-substituted cyclohexylbenzonitriles (**41** and **70–85**) bearing a chain of fixed length (five units) is shown.[86,107–110] Of the series of ethers (**70–73**) differing only in the position of the oxygen atom, only the methoxypropyl-substituted benzonitrile (**73**) with a non-conjugated oxygen atom at some distance from the core exhibits a nematic phase.[107] However, the melting point is higher than that of the corresponding *trans*-4-(4-cyanophenyl)-4-pentylcyclohexane (**41**). Therefore, the temperature range of the nematic phase is much narrower. The angle made by the carbon–oxygen–carbon bond (CH$_2$OCH$_2$) is very similar to that with a methylene unit (CH$_2$) in place of the oxygen atom (CH$_2$CH$_2$CH$_2$). Therefore, it may be assumed that the effect on the clearing point of the oxygen atom in a non-conjugated position in a terminal chain of a particular compound is of a polar and not a steric nature.

Significant variations in the magnitude of the physical properties, such as the liquid crystal transition temperatures, viscosity, birefringence and the elastic and dielectric constants, of alkenyl-substituted compounds with a carbon–

Table 3.7 *Transition temperatures (°C) for the compounds (41 and 70–85)*

R—(cyclohexyl)—(phenyl)—CN

	R	Cr		N		I	Ref
41	⌇	•	30	•	55	•	86
70	⌇O	•	60		–	•	107
71	⌇O	•	23		–	•	107
72	O⌇	•	42		–	•	107
73	O⌇	•	52	•	55	•	107
74	(E) ⌇	•	16	•	59	•	108
75	(Z) ⌇	•	–	[•	−144]	•	108
76	(E) ⌇	•	16	[•	−67]	•	108
77	(Z) ⌇	•	−8	[•	−54]	•	108
78	(E) ⌇	•	60	•	74	•	108
79	(Z) ⌇	•	33	(•	−14)	•	108
80	⌇	•	30	(•	10)	•	108
81	⌇	•	86	[•	1.5]	•	107
82	(E) ⌇O	•	84			•	109
83	(E) ⌇O	•	66	(•	59)	•	109
84	⌇O	•	56		–	•	110
85	(E) ⌇O	•	128	(•	119)	•	110

Parentheses represent a monotropic transition temperature.
Brackets represent a virtual (extrapolated) transition temperature.

carbon double bond (C=C) in the terminal chain, as exemplified by the *trans*-4-alkenylcyclohexyl-4'-benzonitriles (74–80), have generally been found, depending on the position and configuration of the carbon–carbon double bond.[108] This discovery was of significant importance for the successful development of nematic mixtures for super twisted nematic liquid crystal displays (STN-LCDs), see Section 6.[111-113] The results of molecular modelling of ensembles of molecules suggest that steric effects due to the different configuration and, therefore, conformation, of the terminal chain of these alkenyl-substituted compounds are primarily responsible for the very large differences in their transition temperatures and other physical properties.[113] The corresponding acetylene (**81**) derivative[107] exhibits a virtual nematic clearing point just above 0°C. This may be due to the non-axial position of the acetylene group, which would serve to increase the breadth of the molecule, thereby reducing the effective length-to-breadth ratio.

The unexpected and non-linear effect of combining two functional groups, *i.e.* one exerting polar, the other steric effects, on the mesomorphic behaviour of compounds[109] is illustrated to some extent by the transition temperatures of the ether isomers (**82** and **83**) differing in the position of the oxygen atom and the carbon–carbon double bond, see Table 3.7. The ether (**82**) is not liquid crystalline, whereas the other ether (**83**) exhibits a monotropic nematic phase at a temperature higher than that of *trans*-1-(4-cyanophenyl)-4-pentylcyclohexane (**41**). The ester (**84**)[110] is also not liquid crystalline. However, the related compound (**85**) containing an ester group (COO) and an additional carbon–carbon double bond exhibits a very high nematic clearing point, especially for a compound with only two six-membered rings.[110] This is probably a function of an advantageous combination of molecular rigidity and extended conjugation in spite of the presence of the flexible *trans*-1,4-disubstituted cyclohexane ring.[114]

Nematic Materials for Multiplex Addressing

The requirement for large-area, flat panel displays for high-information-content portable instruments, such as computers, personal organisers and notebooks, could only be met at the time by TN-LCDs with more sophisticated addressing schemes than direct addressing due to the high cost of the many drivers required and the physical impossibility of contacting each and every pixel directly. The way chosen initially to resolve this problem was to adopt TN-LCDs with multiplex addressing, see Chapter 2.

Multiplex addressing involves a series of rows of electrodes on one substrate surface at right angles to a similar series of electrode columns on the second substrate surface, see Chapter 2. The overlapping rows and columns create a pattern of rectangular pixels. A combination of a voltage applied to a row with a voltage applied to a column electrode crossing at a given pixel will activate that pixel, if the two voltages are in phase and they create a voltage larger than the threshold voltage of the nematic mixture in the cell formed by the substrates. There is an upper limit to the number of addressable lines (*N*) with acceptable contrast in a TN-LCD with multiplex addressing given by the equation:[93]

$$\frac{V_{ns}}{V_s} = \sqrt{\frac{\sqrt{N-1}}{\sqrt{N+1}}} \tag{18}$$

where V_s and V_{ns} are the select and non-select voltages, respectively. Therefore, as the number of lines to be addressed increases, the more the ratio of the select voltage and the non-select voltages approaches unity and the smaller the acceptable difference between them becomes. Even for a relatively modest degree of multiplexing, *e.g.* for a high-information calculator with 64 addressable lines, there is only a small difference ($\approx 11\%$) between the select and non-select voltages. The contrast also decreases with increasing multiplex rates due to the unintentional activation of adjacent pixels (cross-talk), partly due to fringing field effects. Therefore, the need to address a large number of lines with an acceptable contrast in a TN-LCD with multiplex addressing imposes severe limitations on a nematic mixture compatible with these requirements. It is essential that in the electro-optical response curve, transmission against applied voltage is very steep, *i.e.* a small increase in voltage just above the threshold voltage should lead to a large increase in the transmission. Unfortunately, a consequence of a steep electro-optical characteristic is that the capacity for grey-scale and, therefore, full colour with an array of RGB filters, is severely reduced, since a small voltage change results in a large change in transmission. Therefore, TN-LCDs with multiplex addressing are only capable of displaying a limited amount of black-on-white information with acceptable contrast and brightness. In practical devices this sets an upper limit for the duty cycle of 1:64, *i.e.* only 64 rows can be addressed in one frame time. A duty cycle of 1:32 is common for large, sophisticated calculators.

Nematic mixtures designed for use in TN-LCDs with multiplex addressing had to satisfy a stringent set of specifications including a high $\Delta\varepsilon$ for a low threshold voltage, see Equation 13, a low viscosity for short response times, see Equation 14, and a combination of a low k_{33}/k_{11}, high k_{33}/k_{22} and low $\Delta\varepsilon/\varepsilon_\perp$ for a steep electro-optical response curve. A series of nematic mixtures with these initial properties and a range of values for the birefringence, Δn, were required for displays with different cell gaps driven in the first or second transmission minimum, see Equations 11 and 12. Furthermore, since the transmission curve also depends on temperature, a nematic mixture with a low rate of change in the threshold voltage with temperature ($dV/dT < 0.01 \text{ V} °\text{C}^{-1}$) would eliminate the need to compensate for this change electronically. However, in practice the variation of operating voltages with temperature is often compensated for electronically or by using an optically active dopant whose pitch decreases with increasing temperature. Therefore, the physical properties of existing nematic materials were re-evaluated to assess their suitability for TN-LCDs with multiplex addressing. However, the design and synthesis of new classes of nematic liquid crystals for use in these applications was required in order to optimise the performance of multiplexed TN-LCDs with moderately high-information-content.

The nematic liquid crystals collated in Tables 3.4–3.7 exhibit strong antiparallel molecular correlation,[115-117] probably due to dipole–dipole interactions of neighbouring cyano groups attached to the phenyl ring which have a large dipole moment (4.0 D).[115-119] Thus, the nematic phase formed by rod-like compounds with a strongly polar group in a terminal position can be envisaged as being comprised of a binary mixture of uncorrelated molecules (monomers) and pairs of associated molecules (dimers)[119] in a dynamic, temperature-dependent equilibrium. The effective resultant dipole moment and dielectric anisotropy of the molecular dimers will be much lower than those of the non-correlated monomers due to the antiparallel correlation of the terminal dipoles in the effective molecular dimer. Consequently, the effective molecular length $(1.4 \times l)$ of the molecular dimer is greater than the molecular length (l) of the monomer for the 4-n-alkyl-4'-cyanobiphenyls.[115] This larger anisotropy in the effective length-to-breadth ratio of the molecular dimer than that of unasso-ciated molecules may well be a decisive factor in the formation of a nematic phase at or just above room temperature for the relatively short 4-n-alkyl-4'-cyanobiphenyl molecules. The degree of the molecular association of polar nematic liquid crystals, described by the Kirkwood Froehlich factor g, can be calculated from macroscopic physical data, such as the dielectric constants, refractive index and molecular dipole moments, determined for the bulk nematic material.[118,119] There is no molecular correlation when $g = 1$ and there is complete correlation of the molecular dipoles when $g = 0$. Thus, knowledge of the Kirkwood Froehlich factor can provide important informa-tion on some bulk physical properties of a nematic phase of relevance for the design of new materials and the formulation of nematic mixtures for TN-LCDs with multiplex addressing.

It was soon found that standard nematic mixtures used in TN-LCDs with direct addressing, such as E7 (E. Merck, Darmstadt, Germany), consisting completely of polar compounds with a cyano group in a terminal position, *i.e.* 4-n-alkyl-4'-cyanobiphenyls, 4-n-alkoxy-4'-cyanobiphenyls and 4-n-alkyl-4''-cyano-p-terphenyls, could not be multiplexed to any significant degree. However, it was found[120] that mixtures of the nitriles (**36–69**) shown in Tables 3.4–3.7 and nematic compounds of low dielectric anisotropy, exemplified by the pentyl homologues of the esters (**86–95**)[49,121–126] shown in Table 3.8, consist of a much lower proportion of associated molecular dimers of polar compounds than that found in mixtures consisting purely of nitriles. The higher proportion of non-associated polar compounds not only gives rise to a high value for the dielectric anisotropy and, therefore, a lower threshold voltage, but also a low value of k_{33}/k_{11}. Thus, it was found that mixtures of apolar and polar nematic liquid crystals were much more suited to TN-LCDs with multiplexed addressing than mixtures consisting only of polar liquid crystals with terminal substituents with large dipole moments, *e.g.* the cyano group.

The aromatic phenyl benzoate esters (**86–88**) exhibit a high-viscosity nematic phase with either a low nematic clearing point or a narrow temperature range.[121–123] The phenyl cyclohexyl esters (**89–91**) possess a nematic phase with a relatively wide temperature range and an acceptable viscosity.[121,123] Although the ester (**92**) with two cyclohexane rings only exhibits an enantiotropic smectic B phase, homologues with shorter chains also possess a nematic phase with a relatively high clearing point, low viscosity and a very low birefringence.[124] The bicyclo[2.2.2]octane esters (**93–95**) exhibit a nematic phase with a high clearing point, a low k_{33}/k_{11} ratio, but unfortunately also a high viscosity.[49,125,126] The

Table 3.8 *Transition temperatures (°C) for the esters (86–95)*

Molecular structure		Cr	SmB	N		I	Ref
86		● 36	–	(●	26)	●	121
87		● 50	–	●	58	●	122
88		● 2	–	●	5	●	123
89		● 36	(● 29)	●	48	●	121
90		● 49	–	●	81	●	121
91		● 18	–	●	37	●	123
92		● 51	● 71		–	●	124
93		● 31	–	●	64.5	●	49, 125
94		● 42	–	●	100	●	125
95		● 26	–	●	65	●	49, 126

Parentheses represent a monotropic transition temperature.

presence of a fluorine atom in a lateral position in the esters (**88**, **91** and **95**) leads to a lower melting point, sometimes below room temperature, and the elimination of smectic phases. However, the viscosity of these fluoro-substituted esters is considerably higher than that of the corresponding non-fluoro-substituted esters (**86**, **89** and **93**). Therefore, homologues of the 4-*n*-alkylphenyl *trans*-4-*n*-alkylcyclohexanoates and the corresponding 4-*n*-alkoxyphenyl *trans*-4-*n*-alkylcyclohexanoates[121] were used in combination with nitriles, such as the cyanobiphenyls, to create nematic mixtures for multiplexed TN-LCDs. However, these mixtures exhibited unacceptably long response times, even in low-information-content TN-LCDs, due to the high viscosity of the mixtures, see Equations 14 and 15. Therefore, in attempts to produce apolar nematogens of lower viscosity, a series of new materials (96–105)[127–136] incorporating one or two *trans*-1,4-disubstituted cyclohexane rings with a variety of different central linkages, were synthesised, see Table 3.9.

Most of the compounds (96–105)[127–136] collated in Table 3.9 exhibit a nematic phase as well as an unwanted smectic B phase. However, only the compounds linked directly, *e.g.* (96–99, 102 and 103),[128,129,134] or linked by the ethane linkage, *e.g.* (104 and 105),[135] were found to exhibit the desired low values of viscosity. Apolar two-ring compounds such as (96), which are not mesomorphic at room temperature but exhibit a very low viscosity, and the corresponding three- and four-ring apolar materials, such as compounds (97–99), with very high nematic clearing points are used in commercial mixtures with

Table 3.9 *Transition temperatures (°C) for the apolar compounds (96–105)*

	Molecular structure	Cr		SmB		N		I	Ref
96	C$_3$H$_7$—◯—◯—C$_2$H$_5$	•	1		–	[•	– 70]	•	128, 129
97	C$_5$H$_{11}$—◯—◯—◯—C$_5$H$_{11}$	•	13	•	164	•	170	•	128, 129
98	C$_5$H$_{11}$—◯—◯—◯—C$_2$H$_5$	•	37		–	•	117	•	128, 129
99	C$_5$H$_{11}$—◯—◯—◯—◯—C$_5$H$_{11}$	•	45	•	275*	•	305	•	128, 129
100	C$_3$H$_7$—◯—CO·O—◯—C$_3$H$_7$	•	23		–	•	37	•	124, 130
101	C$_3$H$_7$—◯—O—◯—C$_3$H$_7$	•	7	•	8	•	17.5	•	131
102	C$_3$H$_7$—◯—◯—C$_3$H$_7$	•	64	•	82†		–	•	132
103	C$_3$H$_7$—◯—◯—OC$_2$H$_5$	•	49		–	•	50	•	133, 134
104	C$_5$H$_{11}$—◯—◯—OCH$_3$	•	52	•	72		–	•	135
105	C$_5$H$_{11}$—◯—◯—◯—C$_5$H$_{11}$	•	23	•	135	•	143	•	136

Brackets represent an extrapolated transition temperature; *several ordered smectic phases present.

polar compounds to form multiplexable nematic mixtures with the desired spectrum of physical properties.[128,129] The presence of a fluorine atom in a lateral position in apolar materials, such as (98), leads to the suppression of smectic phases and a more pronounced resultant tendency to form the nematic phase due to steric interactions.[128,129] The suppression of the highly viscous smectic B phase by mixture formulation is usually difficult, especially at low temperatures. Therefore, compounds such as homologues of the cyclohexane ester (100) with longer chains, which only possess a smectic B phase, are not very suitable components of nematic mixtures with a wide nematic temperature range. The presence of a smectic B phase at high temperatures for such compounds can also leads to a limited solubility at low temperatures in nematic mixtures. However, homologues of the cyclohexane ester (100) with shorter chains were used for some time in nematic mixtures of low viscosity and low birefringence.[124] The ether (101)[131] was synthesised in attempts to lower the viscosity even further. However, although this compound exhibits a nematic phase, it is unfortunately of high viscosity. Homologues with longer terminal chains exhibit predominantly smectic B phases.[131] The *trans,trans*-4-*n*-alkyl-4'-*n*-alkylbicyclohexane (102) with no oxygen atom only exhibits smectic B phase, although it was synthesised in the hope that a nematic phase would exhibit a lower viscosity than that of the corresponding ester (100) and ether (101). The corresponding *trans,trans*-4-*n*-alkyl-4'-*n*-alkoxybicyclohexane (103) with an additional oxygen atom exhibits an enantiotropic nematic phase. The two ring ether (104) and the three-ring compound (105) exhibit nematic phases of a viscosity of a somewhat higher magnitude of analogous compounds without an ethyl linking unit.[135,136] However, they can be used to create nematic mixtures with short response times. Homologues with longer alkyl chains are purely smectic. The compounds 96–99 and 103–105 are representatives of important classes of commercial nematic liquid crystals manufactured for use in nematic mixtures for TN-LCDs with multiplex addressing.

Aromatic liquid crystals, such as compounds 106–109 collated in Table 3.10, have been investigated to a much lesser extent since they were perceived to be intrinsically more viscous than the analogous cyclohexane derivatives. This is indeed often the case. It was hoped that the incorporation of a fluorine atom in a lateral position of a polar compound, such as the ester (106)[137], would create a nematic material, such as the esters (107 and 108),[117] of positive dielectric anisotropy and low ratio of k_{33}/k_{11} and $\Delta\varepsilon/\varepsilon_\perp$. Although this was partially achieved, see the relatively low value for the elastic constant ratio k_{33}/k_{11} obtained for the ester (107), this does not improve the multiplexability of mixtures containing such components. Furthermore, the viscosity of these fully aromatic esters is very high. However, fully aromatic apolar compounds, such as 5-*n*-alkyl-2-(4-*n*-alkoxyphenyl)pyrimidines, *e.g.* (109),[138,139] were found to possess very low k_{33}/k_{11} elastic constant ratios, a high birefringence and a low viscosity. Mixed with polar nematics, such as the compounds shown in Tables 3.4–3.7, *e.g.* homologues of compound (41), and related three-ring materials, fully aromatic apolar compounds are important components of nematic mixtures of high birefringence for TN-LCDs with multiplexed addressing.

Table 3.10 Transition temperatures (°C), elastic constant ratio* (k_{33}/k_{11}), birefringence* (Δn), dipole moment (μD), Kirkwood–Froehlich factor (g) and dielectric anisotropy ($\Delta\varepsilon$) for the compounds (106–109)

	Molecular structure	Cr	SmA	N	I	k_{33}/k_{11}	Δn	μ	g	$\Delta\varepsilon$	Ref
106	C₇H₁₅...CN	• 44	—	• 57	•	1.52	0.15	5.6	0.7	19.9	117
107	C₇H₁₅...F CN	• 47	—	• 54	•	1.41		4.8	0.8	9.8	117
108	C₇H₁₅...F CN	• 28	—	• 28.5	•	1.71[a]	0.14[a]	6.1	1.0	48.9	117
109	C₅H₁₁...OC₈H₁₇	• 34	• 56	• 64	•	1.0			1.0	≈0	139

*Measured at 0.95 × T_{N-I}.

Nematic Materials for Active Matrix Addressing

The low contrast and brightness, long response times as well as the strong viewing angle dependency of TN-LCDs with even a moderate degree of multiplexing renders them unsuitable for applications where fast, high-information-content displays are required, such as portable computers and computer monitors. These applications presently use STN-LCDs with a high degree of multiplexed addressing, see Section 6, or TN-LCDs with active matrix addressing, see Chapter 2. TN-LCDs with active-matrix addressing exhibit a relatively high contrast ratio, minimal cross-talk, grey-scale, fast response times and can be fabricated over a relatively large-area (12″–14″ diagonal). They also suffer from either low luminosity or high power consumption, which are a direct consequence of the low intrinsic brightness of the TN-LCD. The cost of the silicon substrate and low production yield due to pixel damage and subsequent repair led initially to high production costs, which retarded the market acceptance of active-matrix TN-LCDs. However, the price differential with comparative STN-LCDs has been reduced substantially by the development of improved production equipment and processes. The market share of TN-LCDs with active matrix addressing of the portable computer displays market is dominant (> 90%).

Most TN-LCDs with active matrix addressing use a silicon substrate with thin-film transistors (TFT) as the active drive element at each pixel. The TFT in a TFT-TN-LCD maintains the charge on the pixel capacitor until the pixel is addressed again in the following frame, *i.e.* it is a memory effect dependent upon the maintenance of the pixel capacitance over time. The loss of charge from the pixel effectively decreases the voltage applied across the pixel, which degrades the contrast. Therefore, it is essential that the resistivity of the nematic liquid crystal mixture should be as constant as possible over time. This requires nematic mixtures with an initial resistivity ($\sigma \approx 10^{12}$–10^{14} Ω cm) several orders of magnitude higher than that of commercial nematic mixtures for standard TN-LCDs with direct or multiplex addressing. The holding ratio, *HR*, is an important factor representing the change in the initial resistivity of the liquid crystal mixture over time:

$$HR = \left[\left(1 - e^{\frac{-2T}{\tau}}\right)\left(\frac{\tau}{2T}\right)\right]^{1/2} \qquad (19)$$

where T is the frame time and τ is the time constant for the pixel and storage capacitor.

The requirement of a very high and constant resistivity over time of nematic mixtures for TN-LCDs with active matrix addressing meant that new liquid crystals which met this and the other specifications, such as low viscosity and high positive dielectric anisotropy, were needed. The nitriles used in nematic mixtures for TN-LCDs with direct or multiplex addressing were soon found to be unsuitable for TN-LCDs with active matrix addressing. The polar nature of the cyano group leads to the solvation of ions from some of the layers on the

LCD substrate. The presence of these mobile ions reduces the resistivity and, consequently, the holding ratio. Therefore, liquid crystals containing substituents with a much smaller permanent dipole moment than that of a cyano group, *e.g.* halogen atoms, especially fluorine atoms, were synthesised, see Tables 3.11 and 3.12, for some representative examples (**110–128**)[140–154] of nematogens synthesised for TN-LCDs with active matrix addressing, although very many related derivatives have been reported.[88,106,109,129,155–157] The corresponding two-ring analogues, which are usually non-mesomorphic, to the three-ring compounds collated in Tables 3.11 and 3.12 are also used in the same mixtures.

The compounds **111–113** with a halogen atom in a terminal position, collated in Table 3.11, exhibit a moderately high value of $\Delta\varepsilon$, which increases with the polarisability of the halogen–carbon bond.[140] The largest value of $\Delta\varepsilon$ exhibited by the halogenated compounds, *i.e.* **113**, is almost half that of the corresponding nitrile (**114**).[141] The viscosity also increases with the size of the terminal halogen atom.[140] Therefore, taking these factors into account, terminally fluorinated compounds appeared to offer the best combination of a moderately high positive value of $\Delta\varepsilon$ and a moderately low viscosity. However, the low value of the dielectric anisotropy of mono-fluorinated nematic liquid crystals, *e.g.* compound **111** in Table 3.11, resulted in unacceptable high threshold voltages. Significantly higher values of dielectric anisotropy were achieved for compounds incorporating terminal groups containing several fluorine atoms bonded to the same carbon atom, *e.g.* compounds **115–118**, incorporating trifluoromethoxy, trifluoromethyl or difluoromethoxy groups, see Table 3.12.[142–147] The nematic clearing point and $\Delta\varepsilon$ generally increase with increasing polarity of the terminal substituent, *e.g.* the trifluoromethyl-substituted compound (**117**) exhibits a larger $\Delta\varepsilon$ value than that of the corresponding nitrile (**114**). A very large $\Delta\varepsilon$ is observed for the sulfonyl fluoride (**118**) due to the high polarisability of the sulfonyl group. Therefore, these and related compounds are found in commercial nematic mixtures for TN-LCDs with active matrix addressing.

Compounds containing two or more fluorine atoms spread around the molecule can exhibit high values of the dielectric anisotropy if the resultant dipole moments of the carbon–fluorine bonds are additive, *c.f.* the dielectric data for the poly-fluorinated compounds (**119–124**) in Table 3.12.[143,148–151] However, the viscosity is not proportional to the number of fluorine atoms in a lateral position, since the molecular rotation volume is not increased by the presence of each fluorine atom. However, the nematic clearing point does decrease with increasing number of lateral fluorine atoms, *e.g.* compounds **123** and **124** with four and five fluorine atoms in terminal and lateral positions, respectively, are not mesomorphic. However, such poly-fluorinated compounds are used in small amounts as components of nematic mixtures in order to increase the positive value of the dielectric anisotropy. Related compounds, which combine a terminal group containing halogen atoms with additional fluorine atoms in lateral positions, *e.g.* compounds **125** and **126** also listed in Table 3.12, can also exhibit values of $\Delta\varepsilon$ comparable to that of the analogous aromatic nitriles.[152] The presence of other electronegative heteroatoms, such as

Table 3.11 Transition temperatures (°C), C–X bond length (Å), dipole moment of C–X bond (Debye) birefringence (20°C), dielectric anisotropy (extrapolated to 100% at 20°C), and viscosity (cP at 20°C) for the compounds (110–114)

	Molecular structure	Cr	SmA	N	I	C–X	μ	Δn	$\Delta\varepsilon$	η	Ref
110	C$_5$H$_{11}$—(cyclohexyl)—CH$_2$CH$_2$—(phenyl)—(phenyl)—H	• 67	—	• 82	•	1.10	≈0	0.15	≈0	22.7	140
111	C$_5$H$_{11}$—(cyclohexyl)—CH$_2$CH$_2$—(phenyl)—(phenyl)—F	• 76	—	• 125	•	1.39	1.47	0.17	+4.9	25.1	140
112	C$_5$H$_{11}$—(cyclohexyl)—CH$_2$CH$_2$—(phenyl)—(phenyl)—Cl	• 100	—	• 158	•	1.78	1.57	0.22	+7.5	46.6	140
113	C$_5$H$_{11}$—(cyclohexyl)—CH$_2$CH$_2$—(phenyl)—(phenyl)—Br	• 125	—	• 163	•	1.93	1.59	0.24	+7.6	63.0	140
114	C$_5$H$_{11}$—(cyclohexyl)—CH$_2$CH$_2$—(phenyl)—(phenyl)—CN	• 79	• 87	• 184	•	1.14	4.0	0.23	+13		141

Table 3.12 *Transition temperatures (°C) and dielectric anisotropy (Δε measured at 20°C; *measured at $T_{N-I} - 10°C$) for compounds (115–128)*

	Molecular structure	Cr	SmB	N	I	Δε	Ref
115	C$_5$H$_{11}$—⟨ ⟩—⟨ ⟩—⟨ ⟩—OCF$_3$	• 43 •		128 • 147	•	+9	143
116	C$_5$H$_{11}$—⟨ ⟩—⟨ ⟩—⟨ ⟩—CF$_3$	• 123	−	• 124	•	+13	145
117	C$_5$H$_{11}$—⟨ ⟩—⟨ ⟩—⟨ ⟩—OCHF$_2$	• 67 •		120 • 162	•	+9	144
118	C$_3$H$_7$—⟨ ⟩—⟨ ⟩—⟨ ⟩—SO$_2$F	• 156	−	[• 60]	•	+27.1	147
119	C$_3$H$_7$—⟨ ⟩—⟨ ⟩—⟨ ⟩—F	• 90	−	• 158	•	+7.3	142
120	C$_3$H$_7$—⟨ ⟩—⟨ ⟩—⟨ ⟩(F)—F	• 46	−	• 124	•	+9.3	143
121	⟍⟍—⟨ ⟩—⟨ ⟩—⟨ ⟩(F)—F	• 48 •		85 • 159	•	+3.2*	151
122	C$_3$H$_7$—⟨ ⟩—⟨ ⟩—⟨ ⟩(F)(F)—F	• 42	−	(• 33)	•	+12.6	149
123	C$_3$H$_7$—⟨ ⟩—⟨ ⟩(F)—⟨ ⟩(F)(F)—F	• 64	−	−	•	+15.2	150
124	C$_3$H$_7$—⟨ ⟩—⟨ ⟩(F)(F)—⟨ ⟩(F)(F)—F	• 123	−	−	•	+20.5	150
125	C$_3$H$_7$—⟨ ⟩—⟨ ⟩—⟨ ⟩(F)(F)—OCH$_2$CF$_3$	• 84	−	• 105	•	+12.2	152
126	C$_3$H$_7$—⟨ ⟩—⟨ ⟩—⟨ ⟩(F)(F)—O—CF=CF$_2$	• 52	−	• 104	•	+12.5	152
127	C$_3$H$_7$—⟨ ⟩—⟨ ⟩—(pyrimidine, F)	• 124	−	• 156	•	+17.7	153
128	⟍⟍—(dioxane)—⟨ ⟩(F)(F)—C≡C—⟨ ⟩(F)—F	• 99	−	• 157	•	+20.8*	154

Parentheses represent a monotropic transition temperature; brackets represent an extrapolated 'virtual' transition temperature; *Sma-N transition at 134°C.

oxygen and nitrogen, incorporated in aromatic or aliphatic rings can also give rise to compounds of strong positive dielectric anisotropy, *e.g.* the pyrimidine (**127**)[153] and the dioxane (**128**).[154]

Many of the compounds of strong positive dielectric anisotropy shown in Tables 3.11 and 3.12 exhibit very high stable resistivity values and solvate ions to a much lesser extent than the analogous nitriles. Therefore, nematic mixtures containing them often exhibit high stable holding ratios as well as low threshold voltages and viscosity. Therefore, due to this advantageous combination of physical properties, many of the polar poly-fluorinated compounds collated in the Tables 3.11 and 3.12 are used in commercial mixtures for TN-LCDs with active addressing.

6 Super Twisted Nematic LCDs[158–173]

It was clear in the early 1980s that the optical performance of highly multiplexed TN-LCDs was completely inadequate, if high-information-content displays for portable instruments, such as notebooks, personal digital assistants and portable computers, were to be realised. TN-LCDs with more than a limited amount of multiplex addressing (1:32 or 1:64), *e.g.* in the first handheld televisions and portable computers, exhibited poor contrast, low brightness and a very strong angle dependency of the viewing angle. Therefore, research was focused on alternative types of LCD. Fortunately, it was found that a range of super twisted nematic (STN) LCDs invented in the mid-1980s had the potential to display large amounts of information with a satisfactory appearance and speed.[158–173] This was due to a very steep increase in the transmission for a small change in voltage above the threshold voltage of LCDs with a highly twisted nematic structure. This allows a very high degree of multiplexing and, therefore, a high information content, according to the formalism of Alt and Pleshko,[174] see Chapter 2. The term STN-LCD is often used to describe the class of LCDs with a highly twisted director configuration as well as one particular type of STN-LCD. Competing technologies based on TN-LCDs with active-matrix addressing, see Section 5, and ferroelectric liquid crystals, which are not discussed here due to space considerations, were still in a development stage.

The steepness of the electro-optical contrast curve of applied voltage against optical transmission increases dramatically for twist angles, Φ, $> 90°$ up to values of $\approx 180°$. Hysteresis and even bistability can be observed for more highly twisted structures ($180° < \Phi > 360°$). However, a stripe texture can also disrupt the appearance of the display under certain conditions.[169] Fortunately, the stripe textures can be suppressed under well defined conditions and bistable displays with very steep electro-optical transmission curves can be produced. The steepness of the electro-optical transmission curve increases with increasing twist angle until a value with infinite slope is found for $\Phi = 270°$, see Figure 3.9. This allows the very high degree of multiplexing essential for high-information-content displays. The first commercial STN-LCDs were addressed using RMS voltages and standard Alt and Pleshko[174] driving schemes. However, although

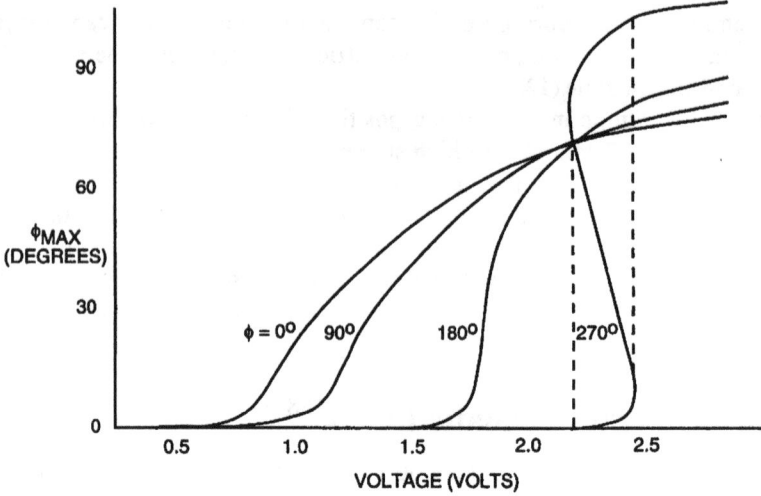

Figure 3.9   *A plot of the twist angle at the centre of a cell against applied voltage for a series of twist angles of nematic LCDs.*[169]

such displays are adequate for high-information-content applications such as notebooks, personal computers, calculators, *etc.*, the relatively long response times of STN-LCDs generate a low contrast at the video addressing rate (< 50 ms) required for television screens or computer monitors. In order to address this problem the so-called Active Addressing™ of STN-LCDs was developed.[175] This electronic drive scheme can produce significantly shorter response times than standard addressing schemes. Therefore, an acceptable contrast for displays at an addressing rate approaching video rate can be realised. STN-LCDs were manufactured in volume shortly after their invention due to the requirement for acceptable flat panel displays with low operating voltages and power consumption for portable instruments, such as calculators and laptop computers. The market share of STN-LCDs in the high-information-content part of the FPD market is being eroded by TN-LCDs with active addressing, see Section 5. However, STN-LCDs are, in turn, steadily displacing TN-LCDs with multiplexed addressing from medium and low-information-content applications, such as calculators, games and personal organisers. STN-LCDs dominate the displays for the personal telephone market.

The first super twisted nematic STN-LCD was reported by researchers from Bell Laboratories, Murray Hill, New Jersey, USA.[158–160] These prototypes and the computer programs used to simulate their optical performance contributed to the discovery of the other commercially successful STN-LCDs, although LCDs based on this concept were not manufactured due to several disadvantageous features described below. A nematic mixture of positive dielectric anisotropy containing a small amount of chiral dopant is homogeneously aligned with a very high pretilt angle ($\theta \leq 45°$) in a glass cell with a large gap ($d < 20\,\mu\text{m}$) between crossed polarisers. The front polariser is aligned parallel to the nematic director in the off-state, which is tilted at a high angle from the substrate surface at both substrates, whereby the molecules of the chiral nematic phase are essentially parallel from the top to the bottom of the cell. There is no twisted structure in the cell. There is extinction between crossed polarisers and the non-addressed parts of the display appear black. If a short pulse of voltage, *e.g.* 2 V, is applied above the threshold voltage, the addressed pixel reverts to the original off-state orientation. However, if a higher voltage pulse, *e.g.* 3 V, is applied the nematic director twists through 360° at various angles to the cell normal to form a second stable state, the on-state. In this orientation two elliptically polarised modes are formed and some light traverses both the polariser and analyser. Thus, a bistable nematic LCD is produced with negative contrast, *i.e.* a bright image against a black background. The chiral dopant increases the energy barriers between the non-twisted and fully twisted state. Other metastable states, which are also possible, can be eliminated by an appropriate combination of device configuration, addressing scheme and properties of the chiral nematic mixture. There are no visible disclinations in the activated areas.

In principle, this mode of operation creates the possibility of producing high-information-content displays due to the short frame times associated with bistable displays, since they are basically a memory effect and only new information must be changed. Unfortunately, metastable twist states of intermediate twist, which degrade the optical performance of the device, form around dust particles in cells with a cell gap below a certain value ($d < 20\ \mu m$). Therefore, since the response time is proportional to d^2, very long response times are observed (≈ 1 s) for LCDs with a cell gap above this critical value. These optically disruptive metastable twist states also form at the interface with spacers used to generate a uniform cell gap.

A super twisted nematic guest–host (STN-GH) LCD containing a dichroic dye in a nematic mixture with a high degree of twist of the nematic director from the bottom to the top of the cell was then found to exhibit a comparably steep electro-optical response curve, but without the formation of undesirable twist states, see also Section 7 for guest–host LCDs.[161–166] The dichroic dye was used to generate the electro-optical effect and no polarisers were used. The amount of light absorption by the dye depends on the direction of its polarisation axis, which is parallel to the LC director in this type of STN-LCD. In the off-state half the light is transmitted, whereas both the elliptically polarised rays are absorbed in the twisted on-state. This produces a positive contrast, *i.e.* black information on a coloured background. Unfortunately, the contrast between the on-state and the off-state transmission is low, since the thick cells ($d > 20$ mm) used in this type of STN-LCD results in a high degree of optical retardation (Δnd), see Chapter 2.

It was then found at the Royal Signals and Radar Establishment, Malvern, UK (now DERA) that STN-GH-LCDs using one polariser and a nematic liquid crystal of high birefringence or a STN-GH-LCD with no polarisers and a chiral nematic liquid crystal of low birefringence both exhibit a very steep electro-optical transmission curve over a very short range of voltages, see Section 7 for a detailed description of this type of guest–host LCD.[161–166]

The STN-GH-LCD using one polariser displays white information against a coloured background for reflective displays with a very high degree of multiplexing (1:500). However, practical displays operate at lower duty cycles (1:200), since the operating voltages are very sensitive to small inhomogeneities in the cell gap (*e.g.* $\pm 0.1\,\mu m$ over a gap of $9.0\,\mu m$). The temperature dependence of the threshold and operating voltages must be compensated for either through the drive electronics or through the use of optically active dopants, which also reduces the maximum degree of multiplexing possible. However, although a polariser is used in this type of STN-GH-LCD, the viewing angle cone is still very wide and the information content is high. Unfortunately, the switching times of this type of display are long partly as a consequence of the high viscosity of the nematic mixture due to the presence of large amounts of chiral dopant used to induce the high degree of twist in the nematic director, but more importantly the presence of the dichroic dye used to generate the electro-optical effect. These problems were resolved by removing the dichroic dye and by adding two polarisers with unusual configurations to generate an optical contrast by the use of interference in the super birefringent effect (SBE) LCD.

Super Birefringent Effect LCDs[167–169]

The super birefringent effect (SBE-LCD)[167–169] reported by Scheffer and Nehring from Brown Boveri in Baden, Switzerland, uses the optical interference of two normal, elliptically polarised modes of transmitted light generated by a high-tilt, highly-twisted nematic structure viewed between two polarisers set in an unusual way, *i.e.* the input polarisation direction is not parallel to the nematic director at either substrate surface and the polarisers are not crossed at 90°, see Figure 3.10.

Device Configuration

Prototype SBE-LCDs combined a chiral nematic mixture with large twist angle, $\Phi \approx 240°-270°$ with a low ratio of the cell gap to pitch ($d/p \approx 0.75$), a positive

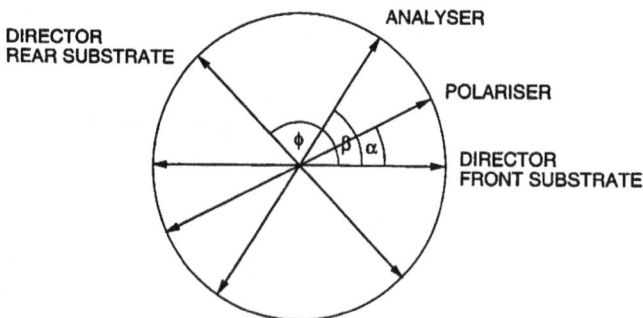

Figure 3.10   *Absorption axes of the polariser and analyse with respect to the director at the front and rear substrate of a STN-LCD.*[169]

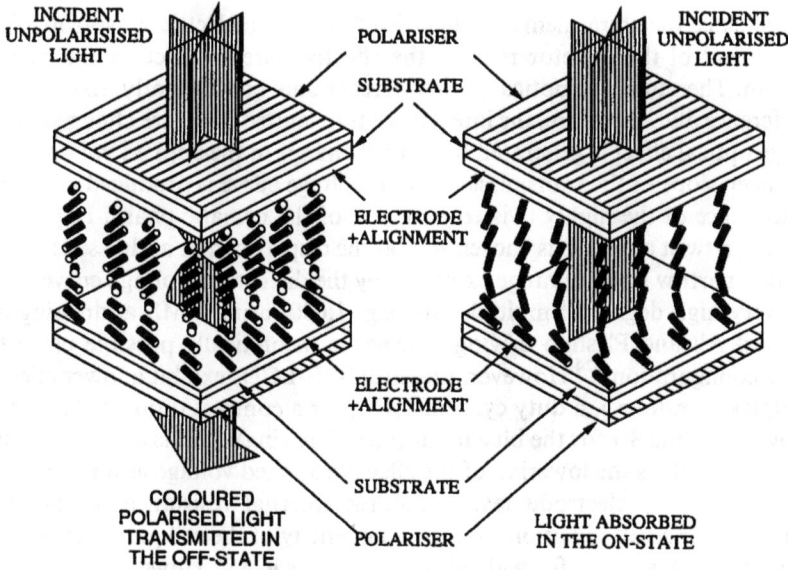

Figure 3.11   *Schematic representation of a super birefringent effect (SBE) LCD.*[167–169]

dielectric anisotropy ($\Delta\varepsilon \approx +12$), a moderately high birefringence ($\Delta n \approx 1.4$) and a very large pretilt angle ($20° < \theta < 30°$), see Figure 3.11. The high pretilt angle was produced using vacuum-deposited SiO/SiO_2. The polarisation axes of the polariser and analyser are set at an angle of α and β, respectively, to the nematic director at the front substrate surfaces, see Figure 3.10.

Off-State

The highly twisted structure of the chiral nematic phase generates two elliptically polarised modes normal to each other. The difference in the optical retardation induced by the chiral nematic phase causes the two rays to be out of phase. The two rays interfere at the analyser; when they are recombined a coloured appearance (blue or yellow) for the static off-state is the result. The highest contrast between the transmission intensity of the off-state and the on-state is produced when the polarisers were positioned at 30° and 60° to the direction of alignment (nematic director) at the front and the rear surfaces, respectively. This combination gives rise to an appearance of positive contrast of black information on a yellow background often referred to as the yellow mode. Rotating one of the polarisers by 90° inverted the contrast to yield white figures on a blue background, referred to as the blue mode.

On-State

Upon the application of an operating voltage just above the threshold voltage results in a modest realignment ($\approx 10\%$) of the nematic director in the middle of the cell since the chiral nematic phase is of positive dielectric anisotropy. This realignment of the director reduces the effective birefringence of the nematic medium. Therefore, the optical path difference and, consequently, the degree of interference and, therefore, the intensity of transmitted light, are changed. Thus, a large optical effect is generated by a small change in the applied voltage and a very steep slope of the transmission *versus* voltage curve is produced as a direct consequence of the highly twisted structure of the nematic phase. In practical displays, a twist angle, ϕ, is chosen so that the display can be addressed at either side of a narrow bistable range exhibited by the distortion–voltage curve.

A very high degree of multiplexing, *e.g.* 1:600, using RMS addressing and standard Alt and Pleshko driving schemes is theoretically possible using this device configuration.[174] However, practical SBE-LCDs exhibit a lower effective multiplexing rate, *e.g.* a duty cycle of 1:120, for a contrast ratio of 10:1 for the yellow mode and 8:1 for the blue mode (direct viewing). This is due to a number of factors, such as the lowering of the effective applied voltage as a result of the resistance of the electrode layers and the internal resistance of the drive electronics, or small variations in the homogeneity of the cell gap. The contrast ratio is modest, *e.g.* 4:1, for wide viewing angles, *e.g.* 45°. However, it is still far superior to that observed for TN-LCDs and GH-STN-LCDs (see Section 7) at a comparable multiplexing ratio.

The response time of prototype SBE-LCDs with normal RMS addressing were long, *e.g.* $\approx$ 300 ms at 20°C for 90% transmission. Therefore, they were not suitable for applications, such as television, where video-rate addressing is essential (response time < 40–50 ms) in spite of a degree of bistability. The highly twisted and tilted configuration of the nematic director in the cell is stabilised by the high tilt angle in both the off-state and in the on-state. The high tilt angle also helps to suppress a competing distorsional structure as well as light-scattering stripe instabilities.[176] It also contributes to a high transmission, I, when the condition

$$\Delta nd \cos^2 \theta_{av} \approx 0.8 \, \mu m \tag{20}$$

is satisfied, where θ_{av} is the average tilt angle in the non-select state. The temperature dependence of the select and non-select voltages must be compensated for either by using the device electronics or by the presence of appropriate optically active dopants in the chiral nematic mixture.[177]

The STN-LCD,[170] independently reported by Kando, Nakagomi and Hasagawa in Japan, is very similar to the SBE-LCD, but with a different polariser combination and lower twist and tilt angle ($\theta \approx 5°$). Tilt angles of this magnitude can be induced using rubbed polyimide, which is a much cheaper and more practical alignment technique than using vacuum-deposited SiO/SiO$_2$, which requires special deposition chambers. This type of STN-LCD contained a chiral nematic mixture of positive dielectric anisotropy with a twist of 180° in the nematic director from the top to the bottom of the cell. The polariser and analyser were parallel with their polarisation axes at angles $\alpha,\beta = 45°$ from the director at the substrates. Thus, the two elliptically polarised rays create a bright yellow background in the off-state. The transmission, I, is given by

$$I \propto \cos^2\left\{\pi\left[1 + \left(\frac{1.00}{\lambda}\right)\right]^2\right\}^{\frac{1}{2}} \tag{21}$$

where the transmission is optimised for yellow light, when the optical path difference

$$\delta = \Delta nd = 1\,\mu\text{m} \tag{22}$$

satisfies this requirement. The application of an appropriate voltage above the threshold voltage gives rise to optical extinction, which produces a positive contrast of black information on a yellow background.

In 1987 researchers at F. Hoffmann-La Roche in Basle, Switzerland reported an optical mode interference (OMI) LCD, whose degree of multiplexability was comparable with that observed for other STN-LCDs.[171–173] However, it had the significant advantage of being an intrinsically black-and-white display of positive contrast with black figures on a white background. Furthermore, its response times were also found to be much shorter than those of comparable STN-LCDs. Moreover, its optical performance was much less sensitive to variations in the cell gap. However, despite this combination of advantageous features, OMI-LCDs suffer from the major disadvantage of poor brightness. Although this can be improved to some extent by increasing the twist angle, the degree of brightness remains substantially lower than that of a comparable STN-LCDs or even TN-LCDs. As a consequence OMI-LCDs have yet to be manufactured on a commercially significant scale.

The major difference between the configuration of the OMI sandwich cell and other STN-LCDs is that the optical path difference ($\delta = \Delta nd \ll 1\,\mu\text{m}$) is much lower. There is no requirement for a significant pretilt ($0 < \theta < 5°$), the twist angle of the chiral nematic layer is lower ($180°$), the front polariser is parallel to the nematic director ($\alpha = 0°$) and the polariser and analyser are crossed ($\beta = 90°$). The $180°$ twist gives rise to strong interference between the two elliptically polarised rays. If the optical path difference is small, *e.g.* $0.4\,\mu\text{m}$, a bright, white, non-dispersive off-state is produced. The chiral nematic mixture should be of positive dielectric anisotropy, low birefringence and exhibit a low cell gap to pitch ratio ($d/p \approx 0.3$).

Electro-optical Performance of STN-LCDs[169,178]

This general description is typical for STN-LCDs with a twist angle $240° < \Phi < 270°$, pretilt angle $5° < \theta < 10°$ and a cell gap, $d \approx 5\,\mu\text{m}$. The steepness of the electro-optical curve, γ, should be as low as possible in order to optimise the number of lines to be addressed:[178]

$$\gamma \propto \frac{\Delta\varepsilon k_{11}}{\varepsilon_\perp k_{33}} \qquad (23)$$

therefore, k_{33}/k_{11} and $\varepsilon_\perp$ should be as large as possible. A large value of $\Delta\varepsilon$ would also lower the threshold voltage. However, since the overall response time is given by

$$t_{on} + t_{off} \propto \frac{\eta\varepsilon_\perp d^2}{\Delta\varepsilon^2}\left(\frac{k_{33}}{k_{11}}\right)^2 \qquad (24)$$

a high value of k_{33}/k_{11} and $\varepsilon_\perp$ result in long response times. Therefore, the easiest way to obtain short response times is to use a cell gap, d, as low as is practically feasible and nematic mixtures with a low viscosity and compromise values for k_{33}/k_{11}, $\Delta\varepsilon$ and $\varepsilon_\perp$ for a steep electro-optical response curve and low threshold voltage. The product $\Delta n d$ is fixed for STN-LCDs operating in the first minimum, e.g. 0.85 for a 240° twist cell. Therefore, Δn should be as high as possible. Unfortunately, the elastic constant and dielectric ratios also determine the upper limit of the ratio of cell thickness to the chiral nematic pitch, before the optically scattering stripe texture is manifested:[176,178]

$$\left(\frac{d}{p}\right)_{max} \propto \frac{\Delta\varepsilon k_{11}}{\varepsilon_\perp k_{33}} \qquad (25)$$

A large k_{33}/k_{11} ratio also permits a greater tolerance of pitch variations for a given cell gap.

Temperature Dependence of Electro-optical Performance of STN-LCDs

The electro-optical characteristics of multiplexed STN-LCDs exhibit a significant dependence on temperature. This has to be compensated in order to avoid variations of the optical performance of the display with temperatures. This can be achieved electronically. However, this problem can also be solved by the use of optically active, chiral dopants. The capacitive threshold voltage of a chiral nematic mixture depends on the pitch of the mixture:[179]

$$V_c^d = V_c^0\left(1 + \frac{d}{p}\frac{4k_{22}\phi_0}{\pi\left(k_{11} + \frac{k_{33} - 2k_{22}}{4}\right)}\right)^{\frac{1}{2}} \qquad (26)$$

where p is the pitch, ϕ_0 the twist angle and d the cell gap. The temperature dependence of the first term of Equation 26 is given by:

$$V_c^0(T) = V_c^0(22\,^{\circ}\text{C}) + \frac{\partial V_c^0}{\partial T}(T - 22\,^{\circ}\text{C}) \tag{27}$$

It is reasonable to assume that the optical threshold voltage is proportional to the capacitive threshold voltage. The twist angle and elastic constant ratios are almost temperature independent. Therefore, since V_c^0 decreases with temperature, then the pitch of the chiral nematic mixture should also decrease proportionately with temperature according to Equation 26, so that V_c^d becomes independent of temperature as possible. Therefore, optically active dopants are used in order to induce the desired twist in the STN-LCD cell, as well as a temperature decreasing pitch. The structure of the chiral dopants should influence other properties of the nematic mixture, *e.g.* the viscosity and the nematic clearing point, as little as possible. Therefore, the chiral dopants should both possess as high a helical twisting power as possible so that only small amounts have to be used. The temperature dependence of STN-LCDs can be successfully compensated for by using two chiral dopants of opposite twist sense.[177]

Black-and-White STN-LCDs

In order to produce black-and-white as well as full-colour STN-LCDs, the monochrome interference colours must first be eliminated. This was achieved initially by using two STN-LCDs in a combined double-layer (DSTN) LCD configuration.[180,181] This involves the use of another non-addressed, passive STN cell in addition to the active display STN-LCD. However, the non-addressed cell has an opposite sense of twist of the nematic director in the cell to that of the addressed STN-LCD. The second STN-LCD, which is identical to the first, but not addressed at all, acts as a retardation compensation layer. The use of an identical second STN-LCD in combination with the active STN-LCD has the advantage that both displays exhibit exactly the same temperature dependence of the birefringence with the same dispersion, assuming that both cells are filled with the same liquid crystal mixture. The second STN-LCD is not addressed and, therefore, there is no increase in power consumption. However, the use of two identical STN-LCDs instead of only one clearly increases the cost and weight of the final product significantly.

Simpler, cheaper and more practical approaches adopted subsequently to compensate for the interference colours of STN-LCDs involve the use of a passive optical element, such as selective polarisers and colour filters, see Chapter 2.[182] A black-on-white appearance is produced most efficiently by using optical retardation layers with a high birefringence and an opposite twist sense to the STN-LCD. However, this also reduces the brightness of the display due to additional light absorption.

Nematic Materials for STN-LCDs

Nematic liquid crystals with a combination of a terminal cyano group and a short alkyl chain were found not only to possess a positive value of the dielectric anisotropy, but also a relatively high ratio of the twist and bend elastic constants, k_{33}/k_{11}.[183–185] This is illustrated by the data collated in Table 3.13 for the nitriles (**36, 41, 53, 39, 49** and **50**).[82,85,86,93,96,99] These nematic compounds combine a positive value of the dielectric anisotropy ($+4.5 < \Delta\varepsilon < +34$) with a strong variation in the dielectric ratio ($0.5 < \Delta\varepsilon/\varepsilon_\perp < 4.3$) and a high bend-to-splay elastic constant ratio ($1.4 < k_{33}/k_{11} < 2.4$), see Table 3.13.[186,187] This combination of physical properties can be used to produce a steep electro-optical characteristic curve and a low threshold voltage in nematic mixtures for use in STN-LCDs. However, these nitriles also induce high values of the rotational and bulk viscosities.[188–190]

Comparisons of values quoted in the literature for the physical properties of liquid crystals are often of dubious validity due to differences in the methods of assessment often carried out at different absolute temperatures (*e.g.* 22°C or 25°C) or reduced temperatures (*e.g.* $T_{N-I} - 10$°C or $0.95 \times T_{N-I}$). The use of extrapolated data from a wide variety of nematic mixtures of different composition and properties at various concentrations is also common. Unfortunately non-ideal behaviour is common for such mixtures and non-linear behaviour is not unusual, *i.e.* the values extrapolated to 100% are more often than not dependent on the matrix used and the concentration of the compound to be evaluated. However, although the absolute values of the data collated in Table 3.13, measured in the same way at the same reduced temperature ($0.96 \times T_{N-I}$), are lower than those reported for the same compounds in the literature,[111,191,192] usually measured at 22°C; the trends and relative values are very similar.

Polar Nematic Materials for STN-LCDs

Mean-field theories,[193,194] and the results of molecular modelling,[113] were used in the design and synthesis of new nematic materials with a high positive value for the dielectric anisotropy and the transverse dielectric constant, low rotational and bulk viscosity and a high k_{33}/k_{11} ratio for use as components in improved nematic mixtures for STN-LCDs. These studies seemed to suggest that the elastic constant ratio k_{33}/k_{11} should be proportional to the length/breadth ratio of individual molecules, *i.e.* long molecules with a narrow rotation volume about the molecular long axis should exhibit a high k_{33}/k_{11} ratio. There is some degree of validity to this theoretical treatment, if ensembles of molecular dimers are taken into account,[113] since molecular dimerisation effectively gives rise to a large length-to-breadth ratio and, hence, in many cases a high ratio of k_{33}/k_{11}.

This interpretation was confirmed to some extent by experimental results, since it had been found that nematic mixtures composed of apolar nematic compounds and polar nematic compounds (nitriles) consist of a number of unassociated polar and non-polar monomers as well as associated polar dimers.[113–120] This not only gives rise to a high value for the observed dielectric anisotropy of the mixture due to the non-associated polar compounds, but also

Table 3.13 Transition temperatures (°C) and some values[187] for the dielectric anisotropy (Δε), the ratio of the dielectric anisotropy and dielectric constant measured parallel to the director (Δε/ε⊥) and the ratio of the bend (k₃₃) and splay (k₁₁) elastic constants for the nitriles 36, 41, 53, 39, 49 and 50

	Molecular structure	Cr	$Sm2$	$Sm1$	N	I	$\Delta\varepsilon$	$\Delta\varepsilon/\varepsilon_\perp$	k_{33}/k_{11}	Ref
36	C_5H_{11}—⟨⟩—CN	• 22.5	—	—	•	• 35	$+13.3^a$	2.0	1.43^a	82
41	C_5H_{11}—⟨⟩—CN	• 31	—	—	•	• 55	$+11.2^a$	2.1	1.77^a	86
53	C_5H_{11}—⟨⟩—CN	• 62	—	—	•	• 100	$+10.0^a$		2.35^a	96
39	C_5H_{11}—⟨⟩—CN	• 71	—	—	(•)	• 52)	$+34.0^b$	4.3	1.13^a	85
49	C_5H_{11}—⟨⟩—CN	• 56	—	—	(•)	• 52)	$+17.4^b$	2.2	1.36^a	93
50	C_5H_{11}—⟨⟩—CN	• 62	• 62	• 43	• 52	85	$+4.5^a$	0.5	1.59^a	99

Parentheses represent a monotropic transition temperature.
ᵃMeasured at a reduced temperature of 0.96 × T_{N–I}.
ᵇExtrapolated value from a 10% concentration in a nematic mixture.

to a low value of k_{33}/k_{11}, since the proportion of molecular dimers with a large effective length-to-breadth ratio is low. Thus, mixing apolar materials of low viscosity, weakly positive or negative dielectric anisotropy and a low k_{33}/k_{11} elastic constant ratio with polar nematic liquid crystals of high positive dielectric anisotropy and a high k_{33}/k_{11} ratio creates nematic mixtures suitable for TN-LCDs with multiplexed addressing, since these require a low ratio of k_{33}/k_{11}, but which are not suitable for STN-LCDs.

Initial attempts to synthesise nematic compounds specifically for STN-LCDs involved the preparation of molecules with polar substituents in lateral positions. The dipole moment of the lateral substituents should have given rise to a high $\varepsilon_\perp$ value, a low $\Delta\varepsilon/\varepsilon_\perp$ ratio and, hopefully, also a high k_{33}/k_{11} elastic constant ratio due to molecular dimerisation, see Table 3.10. However, this was not found to be the case for the phenylbenzoate esters (**106–108**).[117] The degree of molecular association of the 4-cyano-2-fluorophenyl-4-heptylbenzoate (**107**) and 4-cyano-3-fluorophenyl 4-heptylbenzoate (**108**), which both possess a fluorine atom in a lateral position, was found to be much lower than that of the related non-fluoro-substituted 4-cyanophenyl 4-heptylbenzoate (**106**). The complete absence of molecular dimerisation in the nematic phase of the ester (**108**) ($g = 1$) means that the effective length is the molecular length. Therefore, this results in a much lower clearing point for this ester compared to that of the non-laterally substituted ester (**106**) or 2-fluoro-substituted ester (**107**), where dimerisation is present. However, the value found for the k_{33}/k_{11} ratio for the 3-fluoro-substituted benzoate (**108**) is the highest of all three closely structurally related esters. The absence of association combined with the large dipole moment of 6.1 D of the ester (**108**) results in the largest value for the dielectric anisotropy observed for a liquid crystal ($\approx +50$). This fortuitous combination of properties led to the commercial production of the ester (**108**) and other homologues as components of nematic mixtures with a low threshold voltage and operating voltage and a steep electro-optical characteristic for use in STN-LCDs.

A real breakthrough occurred with the synthesis of nematic compounds with an alkenyl substituent, *i.e.* compounds with a non-conjugated carbon–carbon bond in the terminal alkyl or alkoxy chain. A non-conjugated carbon–carbon double bond absorbs in the near UV part of the spectrum well below the cut-off frequency of the glass substrates used in LCDs with non-flexible substrates. Thus, alkenyl analogues of commercial nematic materials with alkyl and alkoxy terminal chains are just as chemically, electrochemically and photochemically stable in the absence of oxygen and moisture. It was found that the physical properties of these alkenyl-substituted material, such as the elastic constants and the nematic clearing point, varied greatly for (otherwise identical) isomers, depending on the position and configuration, *i.e. trans* and *cis* or (*E*) and (*Z*), respectively, of the double bond, see Table 3.14.[111,113,192,195]

Compounds with a *trans* carbon–carbon double bond at an odd number (1 and 3) of carbon atoms away from the molecular core (*e.g.* **74** and **78**), see Table 3.14, exhibit a high nematic clearing point even higher than that of the reference material (**41**)[86] with no double bond in the terminal chain.[195] The analogous *cis*-

Table 3.14 *Transition temperatures (°C), elastic constants (k_{11}, k_{22}, k_{33}, 10^{-12} N), dielectric anisotropy ($\Delta\varepsilon$), dielectric constant measured perpendicular to the molecular long axis ($\varepsilon_\perp$), birefringence (Δn), refractive index measured perpendicular to the director (n_0), rotational viscosity (γ_1, Poise) and bulk viscosity (η, Poise) for trans-1-(4-cyanophenyl)-4-pentylcyclohexane (41), trans-1-(4-cyanophenyl)-4-[(E)-pent-1-enyl]cyclohexane (74) and trans-1-(4-cyanophenyl)-4-[(E)-pent-3-enyl]cyclohexane (78) extrapolated to 100% at 22°C[111–113]*

	41	**74**	**78**
Cr-N	30	16	60
N-I	55	59	74
k_{11}	8.98	9.40	9.41
k_{22}	4.73	6.15	5.74
k_{33}	18.3	22.8	24.1
k_{33}/k_{11}	2.03	2.42	2.56
$\Delta\varepsilon$	12.22	13.03	13.13
$\varepsilon_\perp$	4.85	5.02	5.02
$\Delta\varepsilon/\varepsilon_\perp$	2.52	2.60	2.62
Δn	0.119	0.136	0.130
n_0	1.487	1.493	1.483
γ_1	1.28	1.34	
η	215	221	

isomers (75 and 79) exhibit a very low (extrapolated) clearing point. This has been discussed in detail in Section 5 and will not be dealt with further here. However, it was also found[111–113] that the values of most of the other physical properties of the nitriles (74 and 78)[195] are also equal or superior to those of the reference material (41)[86] with respect to use as components in nematic mixtures for LCDs, especially STN-LCDs, see Table 3.14. The values of the elastic constant ratio k_{33}/k_{11}, birefringence, dielectric anisotropy of the alkenyl compounds (74 and 78) are higher than those of the standard compound (41), whereas the viscosity is comparable. Thus, alkenyl-substituted compounds are used widely as components of nematic mixtures for use in STN-LCDs due to a combination of very steep electro-optical curve, high nematic clearing point, low viscosity and fast response times.

The thermal data collated in Table 3.15 for the compounds (74, 129–133)[195] illustrate the effect of length of the alkenyl chain on the liquid crystal transition temperatures. The nematic clearing point exhibits a clear odd–even effect depending on the number of carbon atoms in the alkenyl chain. The highest clearing points are those of the compounds (130, 74 and 133) with an odd number of units in the chain. This is probably due to a higher degree of anisotropy of the molecular polarisability along the molecular long axis for the compounds 130, 74 and 133 compared to the corresponding materials (129, 131 and 132) with an even number of carbon atoms in the alkenyl chain. The

Table 3.15 *Transition temperatures (°C) for some* trans-*1-alkenyl-4-(4-cyano-phenyl)cyclohexanes (74 and 129–139)*

	Molecular structure	Cr	N	I	Ref
129			56 (• 29) •		108
130 (E)		•	66 • 73 •		108
131 (E)		•	45 55 •		108
74 (E)		•	16 • 59 •		108
132 (E)		•	14 • 39 •		108
133 (E)		•	18 • 49 •		108
134		•	29 • 31 •		108
135		•	50 • 53 •		108
136		•	30 (• 10) •		108
137		•	46 • 53 •		107,108
138		•	19 • 32 •		107,108
139		•	38 • 53 •		107,108

Parentheses represent a monotropic transition temperature.

transition temperatures for the compounds **129** and **134–139**,[107,108,195] also collated in Table 3.15, reveal an odd–even effect for the clearing point and at least one low melting point for a homologue (**138**) with a long alkenyl chain. Alkenyl chains with a terminal carbon–carbon double bond are cheaper to manufacture than alkenyl compounds with a *trans* carbon–carbon double bond due to the absence of stereoisomers. Chemical reactions to form carbon–carbon double bonds normally give rise to a mixture of *trans*- and *cis*-isomers. The *cis*-isomer can be converted by chemical or catalytic procedures into the energetically more favoured *trans*-isomer. However, this is costly and time consuming. Residual amounts of the unwanted isomer still have to be removed, usually by recrystallisation from an appropriate organic solvent. Alkenyl chains with a terminal double bond do not possess *trans*- or *cis*-isomers, since the terminal hydrogen atoms are equivalent. Therefore, they are advantageous from a cost/ performance point of view, although their physical properties are often not as good as those of the corresponding *trans*-isomers with respect to their potential use in mixtures for LCDs.

Apolar Nematic Materials for STN-LCDs

Apolar nematic compounds usually possess a low k_{33}/k_{11} ratio (*e.g.* $1.0 < k_{33}/k_{11} < 1.5$), partially due to the absence of molecular dimers. However, they are still essential components of nematic mixtures for STN-LCDs, since they are used to lower the viscosity and melting point of a nematic mixture of polar components as well as improve the multiplexability of the mixture due in part to the reduction in the proportion of molecular dimers of associated polar molecules. Therefore, the synthesis of the first polar alkenyl liquid crystals with high k_{33}/k_{11} ratios led to the synthesis of a series of apolar alkenyl-substituted compounds with the carbon–carbon double bond in various positions in the terminal chain. Some typical compounds (**140–149**) are shown in Table 3.16.[107,196]

Based on the evaluation of the results of physical measurements on polar alkenyl compounds described above, the position and configuration of the carbon–carbon double bond found in the apolar compounds is similar to that in the polar compounds, essentially for the same reasons. However, the combination of a low melting point and a high nematic clearing point of the bicyclohexane compounds (**142, 143** and **145**) is remarkable. The birefringence and viscosity of these compounds is remarkably low.[107]

The corresponding three-ring compounds, *e.g.* **140–149**, see Table 3.16, exhibit higher melting and clearing points than those of the analogous two-ring materials, as would be expected.[107,196] The low melting points of most of these kinds of alkenyl compounds and the absence of ordered smectic phases, such as the smectic B phase, renders them very useful components of nematic mixtures. They can be used to increase the temperature range of the nematic phase by increasing the clearing point and lowering the melting point of the mixture. The presence of two cyclohexane rings and one phenyl ring gives rise to a low viscosity and a low birefringence for these three-ring materials.[107,108,196] Fluoro- and methyl-substituted compounds are attractive from a synthetic point of view due to the availability of cheap starting materials in an otherwise complex synthetic procedure. The short methyl group is attractive, as a terminal alkyl chain also contributes to a low viscosity and the suppression of smectic phases.

The data collated in the Tables 3.14–3.16 reveals that the presence of a *trans* carbon–carbon double bond in a limited number of specific positions gives rise to a higher clearing point and a higher k_{33}/k_{11} ratio than those of the corresponding material without a double bond in the terminal chain. Smectic phases, especially the ordered smectic B phase, are strongly suppressed. This is

Table 3.16 *Transition temperatures (°C) for the alkenyl-substituted cyclohexane derivatives (140–149)*[107]

	Molecular structure	Cr		SmB		N		I
140		●	42		–	●	58	●
141		●	49		–	●	62	●
142		●	12	●	22	●	62	●
143		●	13		–	●	45	●
144		●	−8	●	52	●	64	●
145		●	19		–	●	32	●
146		●	74		–	●	214	●
147		●	105		–	●	194	●
148		●	87		–	●	186	●
149		●	52	●	104	●	177	●

particularly advantageous for the formulation of nematic mixtures with a low melting point and a high clearing point as well as a steep electro-optical transmission curve. The effect on the nematic clearing point and the suppression of smectic phases is additive to some degree, see Table 3.17.[197,198] Compound 150[197,198] only exhibits a smectic B phase. Compound 151 with one carbon–carbon double bond possesses a smectic B phase at a lower temperature as well as a nematic phase. Compound 152 with two double bonds exhibits a nematic phase with a higher clearing point only than that of compounds 150 and 151. No smectic phase could be observed. The compound 153 with three *trans* carbon–carbon double bonds exhibits the highest clearing point of all and is purely nematic.[198] However, the ratio of k_{33}/k_{11} is only slightly higher for compound 152 with two double bonds than that of the corresponding compound (151) with only one double bond. Thus, the effect of the presence of carbon–carbon double bonds on the elastic constants appears not to be additive, at least not in a linear

Table 3.17 *Transition temperatures (°C) and some values for the viscosity (mPa s) and the ratio of the bend (k_{33}) and splay (k_{11}) elastic constants for the apolar cyclohexane derivatives (150–155)*[197,198]

Molecular structure	Cr	SmB	N	I	γ_1	k_{33}/k_{11}
150	• 36	•	83	-- •		
151	• 37	•	53 •	69 •	27	1.60
152	• 56		- •	90 •	37	1.67
153	• 86		- •	121 •	49	
154	• 39	•	207 •	232 •	381	
155	• 22	•	189 •	253 •	442	1.57

or consistent way. The ratio of k_{33}/k_{11} for the compound with two carbon–carbon double bonds is still much lower than that exhibited by polar materials, such as the phenyl benzoates (106–108), shown in Table 3.10, and the benzonitriles (36, 41, 53, 39, 49 and 50), without a double bond in the terminal chain, listed in Table 3.13. They are very much lower than those of the alkenyl-substituted benzonitriles (74 and 78) shown in Table 3.14.

The viscosity of the three compounds 151–153 increases with the number of carbon–carbon double bonds in the molecule. This is a general phenomenon for alkenyl compounds. The three-ring alkenyl compounds 154 and 155[198] shown in Table 3.17 exhibit higher clearing points than those of the analogous two-ring materials. Although the melting points are low, compounds 154 and 155 exhibit a smectic B phase at very high temperatures. The viscosity is also much higher than that of the analogous two-ring compounds, as expected. The k_{33}/k_{11} ratio is slightly lower than that of the corresponding two-ring materials (151 and 152), perhaps due to a dilution effect attributable to the third cyclohexane ring.

Similar effects have been found for *trans*-4-*n*-alkylcyclohexyl *trans*-4-*n*-alkylcyclohexanoates and the corresponding esters with one or two *trans* carbon–carbon double bonds in the terminal chain.[199–201] However, the ester with two terminal alkenyl chains only possesses a nematic phase over a temperature range of 2°C.[199–201] Related bicyclohexanes with an additional non-conjugated heteroatom in the terminal chain show similar effects, however, they are clearly not additive. This highlights the danger of extrapolating from one mesogenic system to another.[202–204] However, some synergetic effects, due

Table 3.18 *Transition temperatures (°C), elastic constants (k_{11}, k_{22}, k_{33}, 10^{-12} N), dielectric anisotropy ($\Delta\varepsilon$), dielectric constant measured perpendicular to the director ($\varepsilon_\perp$), birefringence (Δn), refractive index measured parallel and perpendicular to the director (n_e and n_o), viscosity (γ_1, $mm^2 s^{-1}$), threshold voltage (V), rate of change of voltage with temperature ($\%°C^{-1}$) and twist angle (°) measured at 20°C unless otherwise stated for three typical nematic mixtures for STN–LCDs[178,206]*

	Low twist STN-LCD[206]	*STN-LCD[178]*	*High twist STN-LCD[206]*
Sm–N	<−30	<−40	<−30
N–I	79	100	71.5
k_{11}	13.1	13.1	
k_{22}	6.9	6.9	
k_{33}	21.1	21.1	
k_{33}/k_{11}	1.61	1.61	
k_{33}/k_{22}	3.1	3.1	
$\Delta\varepsilon$	+24.3	+8.2	+7.39
$\varepsilon_\perp$	6.5	3.9	3.95
$\Delta\varepsilon/\varepsilon_\perp$	3.73	2.10	1.9
Δn	0.15	0.12	0.16
n_e	1.65	1.61	1.67
n_o	1.50	1.49	1.51
γ_1 (20°C)	28	23	16.5
γ_1 (0°C)	128	80	
γ_1 (−40°C)	7100		
V_{10}	1.04	2.15	2.14
dV/dT	0.33		
Twist	180	220	270

to the presence of both an oxygen atom and a carbon–carbon double bond, can give rise to a wide nematic temperature range and a low melting point.[202–204] These effects and the relationship between molecular structure and elastic constants are not well understood.

Alkenyl compounds, such as those shown in Tables 3.14–3.17 are used as major components of nematic mixtures for STN-LCDs with high information content, see Table 3.18.[151,178,205,206] The data are quoted in the table to give some indication of the general magnitude of the physical parameters of mixtures designed for STN-LCDs with different twist angles. Nematic mixtures with a sufficiently high k_{33}/k_{11} ratio can be prepared by adding components with very high k_{33}/k_{11} ratio, such as those shown in Table 3.17, to other, much cheaper, components in order to minimise the final cost of the mixture.[207–210] Chemical companies offer a whole range of nematic mixtures for STN-LCDs. However, these are usually prepared specifically for individual customers and data quoted in catalogues generally only serve as an indication of general properties, which are generally far superior to those quoted.

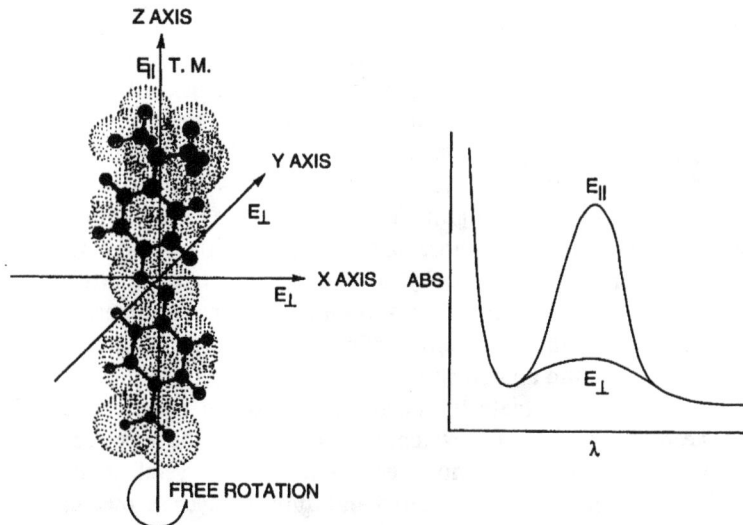

Figure 3.12 *Molecular structure and absorption spectrum of an organic dye with a positive dichroic ratio.*

7 Guest–Host LCDS[211–224]

Guest–host (GH) LCDs are coloured displays,[211–224] whereby the colour and changes in colour are attributable to the absorption of incident light by a dichroic guest dye dissolved in a liquid crystalline host material, usually nematic, and the co-operative reorientation of these dichroic dyes in an electric field, *i.e.* electro-optical devices.

A dichroic dye is a coloured organic compound whose optical absorption depends strongly on the angle of incidence plane polarised light makes with respect to the molecular long axis of the molecule,[225–236] see Figures 3.12 and 3.13. The absorption and emission of light is usually generated by a linear array of conjugated π bonds giving rise to a molecular orbital $n–\pi^*$ or $\pi–\pi^*$ transition from the ground state to the excited state. These transitions are usually associated with the formation of charge transfer states with large dipole moments. Therefore, the molecular shape of a dichroic dye is usually cylindrical and often strongly resembles that of a calamitic liquid crystal. If the electrical vector of incident polarised light is parallel to the transition moment of the dichroic dye then some of the incident light is absorbed with a typical absorption peak and bandwidth. Should the electric vector of incident polarised light be orthogonal to the transition moment of the dichroic dye then little of the light is absorbed. If small amounts of such a dichroic dye (guest) is dissolved in a nematic material (host) and the nematic liquid crystal can then be reoriented by an electric or magnetic field, then the dichroic dye molecules will also be realigned co-operatively by nematic host under the influence of the electric field. Thus, a light shutter for coloured light or a coloured liquid crystal display device can be realised.

A compound to be used as a dye in a coloured LCD must fulfil a number of specifications. The dispersion forces of attraction between a guest dye and the host nematic molecules are at a maximum for a dye molecule with a high length-to-breadth ratio.[235] Furthermore, the dye should be non-ionic as well as chemically, photochemically and electrochemically stable in order to exhibit an acceptable device lifetime. The dyes should also be available in a range of colours. The colour may differ slightly from one nematic host to another due to interaction between the chromophore and the surrounding liquid crystal. Since dyes normally increase the viscosity of the host nematic mixture, a low concentration is desirable in order to maintain short response times. Therefore, chromophores with a high absorption coefficient, which is sometimes referred to as the optical density, and a good solubility ($\approx$ 1–3%) in the liquid crystal host are required in order to create the deep colouration of an effective display. Solubility is especially critical at low temperatures, where the long aromatic dye molecules generally have a pronounced tendency to crystallise out of the nematic mixture. Therefore, only the azo and quinone dyes seemed capable of satisfying these requirements. Typical structures of dichroic dyes developed for GH-LCDs are listed in Tables 3.19–3.23. Furthermore, the transition moment responsible for the visible colour, *i.e.* the charge transfer absorption band excitation, should be parallel to the molecular long axis for most guest–host LCDs. In order to maximise the contrast between the on-state and the off-state, the transition moment of the dyes of positive dichroism should be well aligned to the nematic director, *i.e.* they should exhibit a high order parameter S_{dye}:

$$S_{dye} = \frac{A_\| - A_\perp}{A_\| + 2A_\perp} = \frac{1}{2}(3\cos^2\theta - 1) \tag{28}$$

where θ is the angle between the nematic director and the transition moment of the dye, $A_\|$ is the absorption of light polarised parallel to the director and $A_\perp$ is the absorption of light polarised perpendicular to the director. Clearly $A_\|$ should be as large as possible and $A_\perp$ should be as small as possible for a dye with the transition moment parallel to the long molecular axis. Such a dye is said to be of positive dichroism. If the transition moment is orthogonal to the molecular long axis $A_\perp$ will be larger than $A_\|$ and the dye order parameter will be negative. The dye is then defined as being of negative dichroism, see Figure 3.13. A slightly different order parameter is defined for dyes of this type:

$$S_{dye} = 2(R' - 1) \tag{29}$$

where

$$R' = \frac{A_\perp \cdot \rho_i \cdot n_o}{A_i \cdot \rho_n \cdot n_i} \tag{30}$$

and $A_\perp$, A_i are the optical densities of the homeotropically aligned sample and the isotropic liquid, respectively, ρ_n and ρ_i are the density of the nematic phase

Table 3.19 *Molecular structure, peak absorption ($\lambda_{max,abs}$ nm), colour and order parameter (S) for the dichroic dyes (156–165)*

	Molecular structure	Colour ($\lambda_{max,abs}$)	S	Ref
156		Red (482)	0.17	211, 212
157		Blue (585)	0.55	211–213
158		Blue (595)	0.71	213
159		(450)	0.37	231
160		(359)	0.62	231
161		(400)	0.64	231
162		Purple (560)	0.71	225, 231
163 (n = 1)		Orange (505)	0.46	
164 (n = 2)		Violet (578)	0.51	222
165 (n = 3)		Blue (613)	0.59	

and the isotropic liquid, respectively and n_o, n_i are the refractive indices of the ordinary ray and the isotropic liquid, respectively.

It was postulated that S_{dye} should be higher than 0.73 in order that a good contrast between the on-state and the off-state could be observed.[231] Although this is higher than that found for the nematic host far from the nematic clearing point ($S_{host} \approx 0.65$), the order parameter of the dye itself is usually higher, probably as a consequence of the greater shape anisotropy of the dye molecules. This may also be attributable in part to a lower degree of thermal fluctuation of the dye molecules or a higher local order induced in the nematic host by the dye.[213] A limiting factor for S_{dye} may be that the non-linear shape could give rise

Table 3.20 *Colour, absorption maximum ($\lambda_{max,abs}$, nm), order parameter (S) and solubility (Sol., wt%) for the anthraquinone dyes (166–169)*

Molecular structure	Colour ($\lambda_{max,abs}$)	S	Sol.	Ref
166	Blue	0.64[a]	<0.5	231, 242
167	Blue	0.64[a]	0.6	231, 242
168	Blue	0.66[b]	<1.0	240, 242
169	Yellow (557)	0.66[b]		240, 242

[a]Measured in E43; [b]measured in E7.

to a small transition moment orthogonal to the major transition moment. This would result in a lower contrast or at least a coloration of both the on-and off-states. A related and often encountered measure of the order of the guest dye in the nematic host is the dichroic ratio (R):

$$R = \frac{1 + 2S_{dye}}{1 - S_{dye}} \tag{31}$$

Table 3.21 *Colour, absorption maximum ($\lambda_{max,abs}$, nm), order parameter (S) and solubility (Sol., wt%) for the anthraquinone dyes (170–176)*

Molecular structure	Colour ($\lambda_{max,abs}$)	S	Sol.	Ref
170	Blue	0.73[a]	0.8	247, 248
171	Blue	0.72[a]	1.6	247, 248
172	Blue	0.73[a]		247, 248
173	Blue	0.74[a]		249
174	Blue (640/598)	0.72[a]		249
175	Blue	0.63[a]	<1.0	249
176	Red (618)	0.66[a]	6.0	249

[a]Measured in E7.

Table 3.22 *Colour, absorption maximum ($\lambda_{max,abs}$, nm), order parameter (S) and solubility (Sol., wt%) for the anthraquinone dyes (177–180)*

	Molecular structure	*Colour ($\lambda_{max,abs}$, nm)*	S	*Sol.*	*Ref*
177		Yellow (465)	0.79[a]	1.7	247, 248
178		Yellow	0.79[a]	9.2	247, 248
179		Orange	0.80[a]		247, 248
180		Red (520)	0.80[a]	15	247, 248

[a]Measured in E7.

Table 3.23 *Colour, absorption maximum ($\lambda_{max,abs}$, nm), order parameter (S), solubility (Sol., wt%) and transition temperatures (°C) for the tetrazine dyes (181[a], 182[b] and 183[c]) and the azulene dyes (184[d] and 185[e])*

Molecular structure	Colour ($\lambda_{max,abs}$)	S	Sol.	Ref
181	Red/violet (550)	0.7	15	252
182	Red			252
183	Red	0.78	< 30	253
184	Red (480)	0.21		254
185	Red			255

[a]Cr–SmC, 49°C; SmC–N, 52.5°C; N–I, 63°C; [b]Cr–SmC, 127°C; SmC–N, 132.5°C; N–I, 180°C; [c]Cr–N, 107°C; N–I, 211°C; [d]Cr–I, 124°C; N–I, 116°C; [e]Cr–S, 190°C; S–I, 250°C.

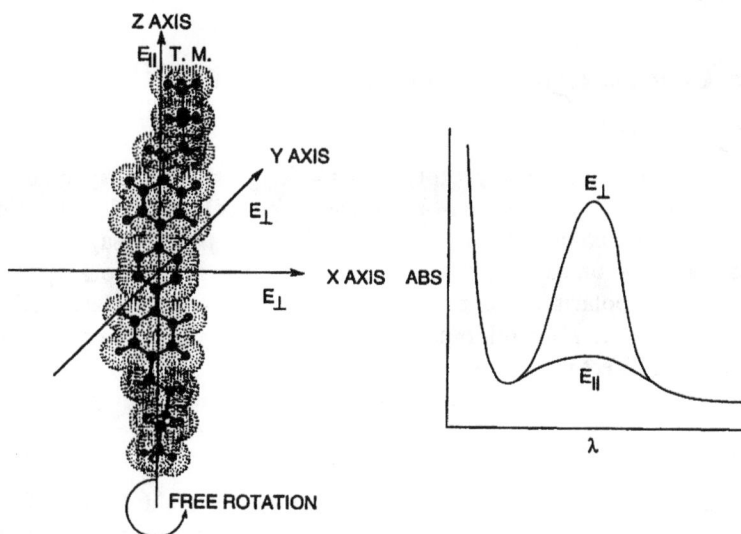

Figure 3.13 *Molecular structure and absorption spectrum of an organic dye of negative dichroic ratio.*

A value of $R > 6$ is generally regarded as necessary for a GH-LCD with an acceptable contrast.

There are many variants of this type of LCD. However, there are two main types: the first uses a dichroic dye, one polariser and a nematic host; the second type utilises a dichroic dye dissolved in a chiral nematic host and does not require any polarisers. Both types can be used with nematic liquid crystals of positive or negative dielectric anisotropy and dichroic dyes with positive or negative dichroism. The first nematic guest–host effect was developed in the late 1960s at the RCA Laboratories[211,212] at the same time as the DSM-LCD, see Section 2. The second kind of GH-LCD was reported[213] six years later by White and Taylor of Bell Laboratories, New Jersey, USA. The original versions of both types of GH-LCDs were of negative contrast, *i.e.* white information against a coloured background. Human factor studies suggest that positive contrast displays, *i.e.* coloured information against a white background, exhibit a higher degree of readability compared to equivalent negative contrast displays with the same intrinsic contrast ratio between addressed and non-addressed areas. The contrast of a given type of LCD can generally be inverted by driving the background instead of the pixellated areas. However, this involves much larger active electrode areas with the associated higher power consumption and the possibility of electric field fringe effects.[214] Heilmeier and White and Taylor GH-LCDs with positive contrast have been developed. However, only White and Taylor GH-LCDs are of any real commercial interest due to the absence of polarisers which are costly, absorb light and limit the already restricted viewing angle further. There is a resurgence of interest in this type of display after more than 25 years due to these very reasons, although a limited number of GH-LCDs have been manufactured during this time for the consumer market, *e.g.* coloured digital watches.

Negative Contrast Heilmeier and Zanoni GH-LCDs[211,212]

Display Configuration

The first nematic guest–host prototype nematic guest–host display device contained a nematic liquid crystal (4-butoxybenzoic acid) and a pleochroic dye (methyl red or indophenol blue) sandwiched between two (Nesa) electrodes ($d \approx 12\,\mu m$) rubbed uniaxially, but with no additional orientation layer, see Figure 3.14. One polariser was fixed to the front substrate surface with its direction of maximum absorption parallel to the rubbing direction and, therefore, the nematic director.

Off-State

The nematic liquid crystal mixture containing the pleochroic dye is of positive dielectric anisotropy and is aligned parallel with the director parallel to the substrate surfaces. Therefore, plane polarised light is absorbed by the dye molecules in non-addressed areas of the display and they appear coloured.

On-State

The application of an electric field between the electrodes results in a realignment of the nematic liquid crystal mixture and the dichroic dye molecules parallel to the electric field resulting in a lower optical density (absorption) and, theoretically, the disappearance of colour assuming an ideal order parameter ($S = 1$) of the nematic liquid crystal director and the dye molecules. A residual absorption in this state gives rise to a display with a strongly coloured background and weakly coloured information.

A Heilmeier and Zanoni GH-LCD with positive contrast has a very similar construction. A suitable surface treatment results in a homeotropic orientation for a host nematic mixture of negative dielectric anisotropy incorporating a dichroic dye of positive dichroic ratio.[216,217] The electric vector of incident polarised light is perpendicular to the absorption moment of the dissolved dye. No light is absorbed and the background appears colourless. An applied electric field above a threshold value realigns the nematic host of negative dielectric anisotropy and the guest dye parallel with the cell substrates. The transition moment of the dye is now parallel with the electric vector of the transmitted polarised light, which is consequently absorbed. Therefore, activated pixels appear coloured against a white background.

Another version of the Heilmeier and Zanoni GH-LCDs[215,216] with positive contrast is essentially the inverse of the Heilmeier and Zanoni GH-LCDs[211,212] with negative contrast described above. A nematic mixture of negative dielectric anisotropy incorporating a dichroic dye of negative contrast is aligned parallel

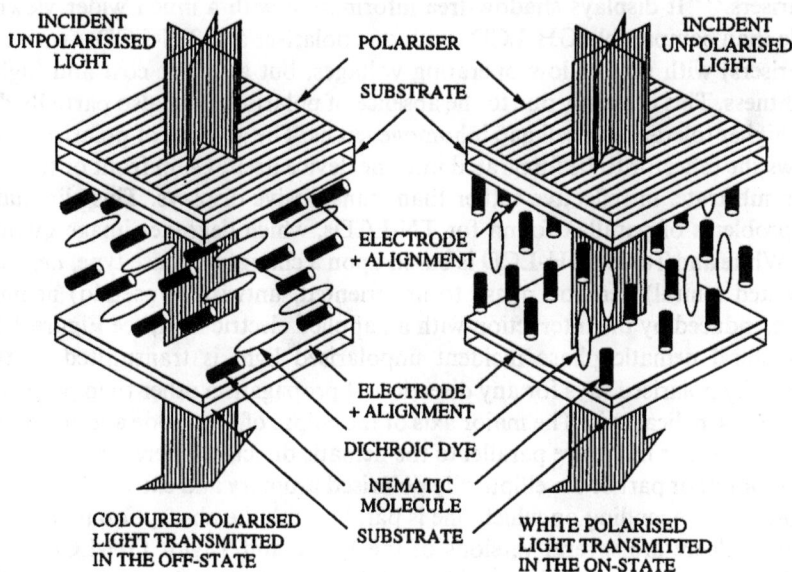

Figure 3.14 *Schematic representation of a Heilmeier and Zanoni guest–host liquid crystal display (GH-LCD).[162,163]*

to the substrate surface in a standard cell with one polariser. Since the transition moment of a dye of negative dichroism is perpendicular to its long molecular axis,[217] see Figure 3.13, and, therefore, also to the direction of propagation of the linearly polarised light, absorption should not take place. Therefore, the non-addressed background pixels appear colourless or weakly coloured. An applied voltage above the threshold value results in the dielectric realignment of the host nematic mixture and the guest dye parallel to the field. The molecular axes of both now lie in the plane of the cell orthogonal to the field. However, the transition moment of the dichroic dye is now parallel with the electric vector of the incident polarised light. The light is absorbed and activated pixels appear coloured against a white background. The contrast ratio of this display mode is lower than other configurations due to the low dichroic ratio (≈ 5) of the known dyes of negative dichroism and the fact that the transition moment rotates rapidly round the long molecular axis and still absorbs some light in the off-state.

White and Taylor GH-LCDs[213]

The GH-LCD reported by White and Taylor in 1974 possesses a number of advantageous features for simple alpha-numeric displays (two electrical con-tacts per pixel segment) compared to the related GH-LCDs originally reported by Heilmeier and Zanoni as well as TN- and STN-LCDs, which utilise polarisers.[213] It displays shadow-free information with a much wider viewing angle than comparable GH-LCDs with one polariser and TN-LCDs (with two polarisers) with similar low operating voltages, but at lower cost and higher brightness. This is mainly due to the absence of polarisers, but also partially due to a higher tolerance of cell gap inhomogeneities. The absence of polarisers also allows the reflector to be integrated into the device itself, *i.e.* in front of the rear glass substrate, in reflective rather than transmissive displays. This eliminates the problems of parallax found for TN-LCDs, which degrades image quality. The White and Taylor GH-LCD is based upon a change of phase type, *i.e.* from a twisted (chiral) nematic phase to an oriented, untwisted (pseudo) nematic phase, induced by the interaction with an applied electric field, see Figure 3.15. In a chiral nematic phase incident unpolarised light is transmitted as two elliptically polarised rays for any direction of propagation other than perpendi-cular to the helical axis. The major axis of the ellipse of one mode and the minor axis of the other mode are parallel to the nematic director. Therefore, either the major or minor part of the elliptically polarised ordinary and extraordinary rays is absorbed, depending on which one is parallel with the absorption moment of the dye. There are several versions of the White and Taylor GH-LCDs with positive and negative contrast, which exhibit important differences, especially with respect to boundary conditions, sign and magnitude of the dielectric anisotropy of the nematic mixture and the transition moment of the guest dye.

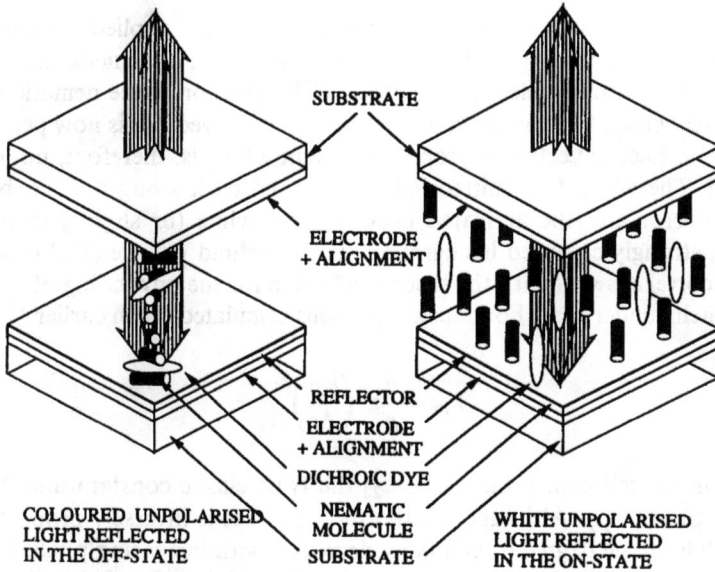

Figure 3.15 *Schematic representation of a White and Taylor guest–host liquid crystal display (GH-LCD).*[213,214]

Negative Contrast White and Taylor GH-LCDs[213]

Display Configuration

The electrode surfaces of a normal LCD sandwich cell ($d \approx 8$–$10\,\mu$m) are coated with an alignment layer in order to induce a planar alignment of a host (chiral) nematic mixture containing the dichroic dye of positive dichroism and a chiral dopant. Due to the absence of polarisers a very thin mirror can be incorporated within the cell on top of the rear glass plate electrode in direct contact with the guest–host mixture, see Figure 3.15.

Off-State

The planar boundary conditions align the molecules of the chiral nematic mixture and the dye molecules in the azimuthal plane, *i.e.* parallel to the device substrates. The twisting power of the chiral dopant then gives rise to the spontaneous formation of a helical structure of pitch, *p*, with an axis perpendicular to the substrate surfaces (Grandjean texture), *i.e.* from the rear substrate to the front substrate. Thus, incoming unpolarised light is absorbed efficiently by the dye, which is distributed through an angle of 360° in the plane of the cell. Thus, non-activated areas of the display appear strongly coloured.

On-State

The application of an electric field above a threshold value orients the director of the nematic guest–host mixture parallel to the field applied between the electrodes, thereby leading to the unwinding of the helix, if the nematic mixture is of strongly positive dielectric anisotropy. The director of the nematic phase and, consequently, the transition moment of the dissolved dye is now perpendicular to the electric vector of propagating light which is, therefore, no longer absorbed. Therefore, transmitted light is reflected back along its path by the internal mirror and the activated pixels appear white (or slightly coloured) against a strongly coloured background. The threshold voltage (V_{th}) of actual displays correlates well with the theoretical value for the unwinding of a chiral nematic helix under these boundary conditions calculated much earlier.[218]

$$V_{th} = \frac{d}{p}\pi^2 \left(\frac{k_{22}}{\Delta\varepsilon}\right)^{\frac{1}{2}} \tag{32}$$

where d is the cell gap, p the pitch, k_{22} the twist elastic constant and $\Delta\varepsilon$ the dielectric anisotropy. Although the threshold voltage is inversely proportional to the pitch, p, the degree of ellipticity decreases with increasing pitch, *i.e.* λ/p should be small. Therefore, since the contrast, C, is highest for almost circularly polarised light, a compromise between threshold voltage and on-state transmission, I_{on}, must be made:

$$C = I_{on}/I_{off} \tag{33}$$

where the background brightness, I_{off}, is a constant for a given display. Since the contrast ratio is found empirically to depend on the product Δnp, a small Δn value allows a larger value for p and, therefore, a lower threshold voltage for a given cell thickness ($d/p \approx 3$ for optimum contrast). Similarly, although increasing the dye concentration also increases the contrast ratio, the brightness is found to decrease proportionately. However, commercial displays exhibit moderate contrast at low threshold voltages in relatively thick cells, *e.g.* 5:1 at 3 V in a 10 μm thick cell, where standard cell gap variations are inconsequential. A turbid, cloudy appearance is produced for significant cell gap inhomogeneity. High brightness can also be achieved with a suitable internal mirror.[214] However, a major disadvantage of this configuration is that high operating voltages are required to align the optic axis parallel to the electric field in order to obtain good contrast. Furthermore, metastable disclination lines formed after a given pixel has been deactivated can give rise to a visible after image.

Positive Contrast White and Taylor GH-LCDs[214,216,219–221]

Display Configuration

A suitable surface treatment results in a homeotropic orientation for a nematic mixture incorporating a dichroic dye of positive contrast and an amount of a chiral dopant insufficient to overcome the surface forces and generate a twisted structure in the nematic phase.

Off-State

The static homeotropically aligned guest–host nematic mixture (plus guest dye and chiral dopant) is optically transparent and, therefore, the display appears colourless in the non-activated state. The lower limit of the pitch for a given cell gap, before a twisted nematic structure becomes energetically more favoured than the homeotropic nematic structure, is determined by the d/p ratio:[221]

$$\frac{d}{p} < \frac{k_{33}}{2k_{22}} \approx 1 \tag{34}$$

where k_{33} is the bend elastic constant. The homeotropic boundary conditions of this GH-LCD should result in the helix axis lying in the plane of the cell. However, the observed texture, referred to as the scroll texture is made up of a multitude of left- and right-handed helices. Indeed the helix axis is found to still lie predominantly perpendicular to the cell walls. However, this texture does not scatter light and replaces the scattering texture responsible for the formation of after-images in cells with parallel boundary conditions, resulting in their elimination for this display configuration.

On-State

The application of an electric field above the threshold value results in a reorientation of the nematic liquid crystal mixture, if the nematic phase is of negative dielectric anisotropy. The optically active dopant then applies a torque to the nematic phase and causes a helical structure to be formed in the plane of the display. The guest dye molecules are also reoriented and, therefore, the display appears coloured in the activated pixels. Thus, a positive contrast display is produced of coloured information against a white background. The threshold voltage is dependent upon the elastic constants, the magnitude of the dielectric anisotropy, and the ratio of the cell gap to the chiral nematic pitch:

$$V_{\text{th}} = 2\pi \sqrt{\left(\pi \frac{k_{33}}{\Delta\varepsilon}\right)} \sqrt{1 - \left(\frac{2k_{22}}{k_{33}}\frac{d}{p}\right)^2} \tag{35}$$

Super Twisted Nematic (STN) GH-LCDs[167–169]

A super-twisted version of the Heilmeier and Zanoni GH-LCD[167–169] with one polariser and a nematic liquid crystal of low birefringence exhibits very steep electro-optical transmission curves and a high degree of multiplexability.[161–166] This is of great advantage compared to both the standard Heilmeier and Zanoni and White and Taylor GH-LCDs due to the ability to display high information content at high contrast and low voltage. This is especially advantageous using achromatic guest–host mixtures in order to produce black-on-white displays.

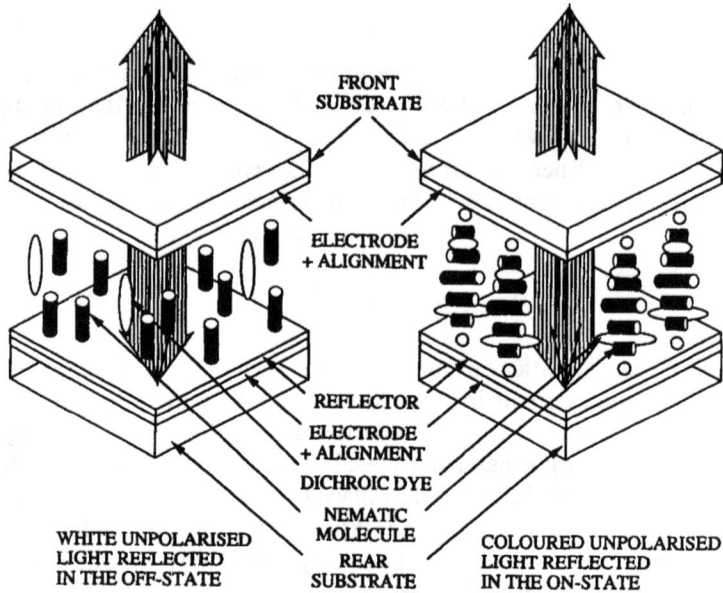

Figure 3.16 *Schematic representation of a super twisted White and Taylor guest–host liquid crystal display (STN-GH-LCD).*[161–166]

Display Configuration

The cell contains a nematic mixture with a twist of 270° and homogeneous alignment with a high pretilt angle (θ), see Figure 3.16. The nematic mixture is composed of one or several dichroic dyes, a chiral dopant and a nematic host of low birefringence.

Off-State

Incident plane polarised light after traversing the polariser is absorbed by the dichroic dye, whose transition moment is parallel to the electric vector of the light. The internal reflector reflects the coloured light back through the polariser. Thus, the non-addressed background appears coloured.

On-State

The application of a surprisingly low voltage, *e.g.* 1 V, results in the reorientation of the nematic host and the dissolved dyes parallel to the applied field and orthogonal to the plane of the cell as the chiral nematic twist is unwound. Therefore, the transition moment of the dye is aligned perpendicular to the electric vector of the plane polarised light and little or no absorption takes place. Thus, activated pixels appear white against a coloured background. The difference in voltage between the threshold voltage, V_{th}, and the voltage, V_{on}, at which the cell can be regarded as activated and light transmission is high, is

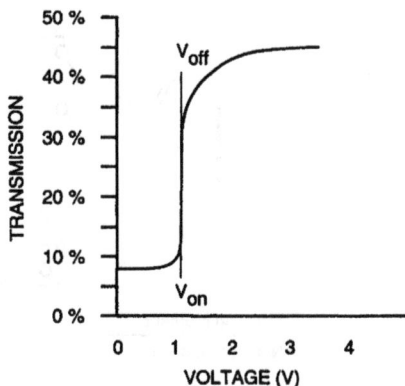

Figure 3.17 *Schematic representation of the transmission against voltage for a guest–host LCD.*

small. This results in a very steep electro-optical characteristic, see Figure 3.17. Therefore, since the number of addressable lines according to Alt and Pleshko is proportional to the ratio V_{on}/V_{th}, a large number of lines can be multiplexed, e.g. 200 for $V_{on}/V_{th} = 1.07$ and $\theta = 30°$. The ratio V_{on}/V_{th} is a constant for $0.5 < d/p < 1.0$. However, since there is a linear dependence of V_{on}/V_{th} on d/p, then p should be as large as possible within the above limit for a given cell gap. The temperature dependence of the operating voltages can be compensated either electronically or by using a chiral dopant whose pitch decreases with temperature.

Dichroic Dyes — Guests

Dichroic dyes of positive and negative dichroism are required for various configurations of the Heilmeier and Zanoni as well as the White and Taylor GH-LCDs with positive and negative contrast. They should be chemically, photochemically and electrochemically stable, as well as exhibit a high order parameter, a high dichroic ratio and good solubility in the host nematic matrix.

Positive Contrast Dyes

The dye used in the original guest–host LCDs with one polariser reported by Heilmeier and Zanoni[211,212] was a standard commercially available dye called methyl red (**156**), shown in Table 3.19. Methyl red is an azo dye with a cylindrical shape, a relatively high length-to-breadth ratio and a transition moment parallel to the molecular long axis. Therefore, it exhibits a relatively high positive dichroic ratio. A number of other commercially available dyes with a less linear molecular structure, such the indophenol dye (**157**), were also investigated.

The first prototype GH-LCDs based on the chiral nematic phase change effect investigated initially by White and Taylor also used indophenol dyes.

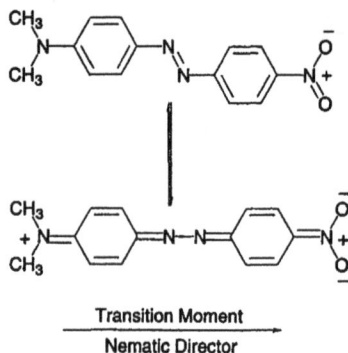

Transition Moment
───────────────────────
Nematic Director

Figure 3.18 *Canonical structures of an azo dye in the ground state and the excited state.*

However, the order parameter of these dyes was found to be too low to generate a high contrast ratio.[213] Therefore, a blue thiazole azo dye (**158**), with a quasi liquid crystalline molecular structure, also shown in Table 3.19, was synthesised specifically for improved chiral nematic phase change GH-LCDs. The presence of an electron-donating and an electron-withdrawing group at each end of the molecule leads to a large energy difference between the n ground state and the π^* excited state. It is this process of absorption, transition from the n to π^* states and subsequent radiative decay, which is generally regarded as responsible for light emission in azo dyes. The excited state is a charge transfer complex, whose transition moment is parallel to the director of the nematic host, see Figure 3.18. The polar nature of asymmetrically substituted azo dyes, such as **158**, in the ground and excited states gives rise to a large dipole moment in the direction of the molecular long axis. This facilitates solvation and orientation in a nematic host of positive dielectric anisotropy and co-operative reorientation an electric field. Such asymmetrical azo dyes with liquid crystalline properties had been synthesised many decades earlier.[238] The use of long thin azo dyes, such as **158**, resulted in a higher brightness and contrast, due to the higher order parameter, than similar prototypes using the indophenol dye (**157**).[211–213] Thiazole azo dyes, such as **158**, had been developed originally for colouring man-made fibres.

It was postulated that the order parameter should increase with increasing length-to-breadth ratio, *i.e.* anisotropy of molecular shape.[229–232,239] This was indeed found to be case for the four azo dyes (**159–162**) and the three merocyanine dyes (**163–165**; n = 1, 2 and 3) collated in Table 3.19.[225] The dichroic ratio of the tri-azo dyes (**161** and **162**) is significantly higher than that of the related mono-azo and di-azo dyes (**159** and **160**). The mono-azo dye (**159**) exhibits a particularly low order parameter. Unfortunately the solubility also decreases with increasing molecular length. Moreover, even with very long thin dyes, such as compound **161**, there is still some absorption in the supposedly colourless state of GH-LCDs. This lowers the observed contrast since both the on-state and the off-state are coloured, strongly and weakly, respectively. However, azo dyes such as those shown in Table 19 appeared more-or-less to satisfy most of the requirements of the guest–host chiral nematic phase change

LCDs. However, accelerated life-time tests revealed that photochemical degradation of many azo dyes in strong sunshine would lead to unacceptably short device lifetimes. This was especially true of azo dyes absorbing in the blue region of the visible spectrum. The fading of the dye due to photochemical degradation in intense sunlight was found to follow first order kinetics, *i.e.* to be linearly dependent on time at constant illumination.[233,234] Since the optical density is proportional to the concentration of the dye according to the Beer–Lambert law, then

$$\log_e \frac{A_0}{A_t} = kt \qquad (36)$$

is found to be valid,[233] where A_0 is the optical density before illumination and A_t is the optical density at time t, and k is a rate constant. Some of the decomposition products collect at the substrate surface and gave rise to an homeotropic alignment of the nematic director and the dissolved dye. This can lead to an apparent acceleration of the fading of the dye. Although the glass used as substrates in GH-LCDs actually removes most UV light up to the glass cut-off point ($\approx 300\,\text{nm}$), azo dyes still absorb in the UV in this region. However, the use of UV filters was found to significantly improve practical display lifetimes by absorbing most of the residual UV radiation above the glass cut-off point. It is this residual UV which is responsible for the photochemical degradation of most dyes with colours from yellow to red. However, GH-LCDs incorporating azo dyes were manufactured for devices which are not exposed to continual sunlight, such as digital watches.[234] However, UV filters do not significantly inhibit the photochemical degradation of blue azo dyes. The rate of decomposition of blue azo dyes is much higher than that of azo dyes of any other colour. This not only prohibits the creation of blue GH-LCDs, it prevents the formulation of achromatic black dye mixtures. Unfortunately, although the incorporation of lateral substituents in azo dyes gives rise to a substantial increase in colour fastness for textile applications, attempts to improve the stability of blue or other azo dyes designed for GH-LCDs by incorporating lateral substituents did not result in longer lifetimes in nematic mixtures.[234] Dyes incorporating lateral substituents were also found to exhibit lower order parameters, although greater solubility, than equivalent non-substituted dyes. This is probably due to the lower degree of molecular shape anisotropy. The effect on the order parameter of multiple lateral substitution was not additive as this effect is probably due to the steric effect of broadening the molecule of the largest lateral substituent.[234]

Similar lifetime issues were encountered for a large variety of dyes with an elongated molecular structure with an extended conjugated core similar to that of the azo dyes, such as the merocyanine dyes (**163–165**; $n = 1$–3), shown in Table 3.19, as well as azomethines and methine dyes.[227,235] However, most of these exhibit much lower order parameters and dichroic ratios than conventional azo dyes prepared earlier.

Anthraquinone dyes were already known to exhibit good thermal stability

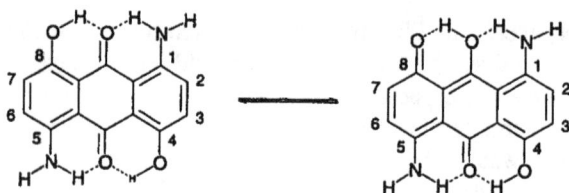

Figure 3.19 *Isomeric structures of an anthraquinone dye used in guest–host LCDs.*

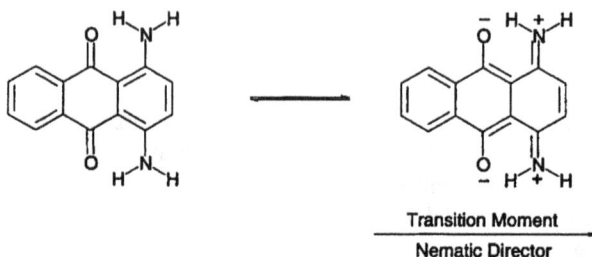

Transition Moment

Nematic Director

Figure 3.20 *Isomeric structures of a diamino-substituted anthraquinone dye.*

and light fastness, *e.g.* for dyeing textiles. This is due in part to hydrogen bonding between the hydrogen atoms on the amino and hydroxyl functions with the adjacent carbonyl groups, see Figure 3.19. They also possess a transition moment parallel to the director of the host nematic mixture, see Figure 3.20. The excited state formed by a π–π^* transition is a charge transfer state with a large dipole moment. Therefore, many anthraquinone dyes[240–251] used in plastics and textiles, such as some of those collated in Tables 3.20–3.22, were investigated as potential dichroic dyes for GH-LCDs. Accelerated sun tests established that anthraquinone dyes,[240] such as **166–169**, were much more resistant to photochemical degradation in nematic host mixtures than the azo dyes investigated previously. The rate constants for fading of the dye under identical illumination conditions are at least an order of magnitude lower for the anthraquinone dyes than those of comparable azo or merocyanine dyes.[233,234] Unfortunately, the order parameter ($S_{dye} \approx 0.64$) and solubility (< 1 wt%) of simple non-substituted anthraquinones, such as **166** and **167**, were found to be too low to obtain a satisfactory contrast.[233–235] Since the optical density of anthraquinone dyes is three or four times lower than that of azo dyes of the same colour, correspondingly higher amounts of the dichroic dye must be dissolved in the nematic host in order to obtain the same deepness of colour. Although the presence of one or two aromatic substitutents on amino functions in positions 1, 4, 5 or 8 in modified anthraquinone dyes such as **168** and **169** led to higher order parameters and solubility,[225,236,240,242] these were still too low for satisfactory operation of GH-LCDs. Higher order parameters were obtained for dyes such as **170–176** shown in Table 3.21, with one alkyl or aryl substituent in position 2 or two alkyl or aryl groups in positions 6 or 7 adjacent to the amino and hydroxy functional groups in the anthraquinone core.[225,236,247–249] Unfortunately, the solubility and range of colour of these dyes were also found to be insufficient for satisfactory display performance, especially for black-and-white displays. However, anthraquinone dyes incorporating sulfur atoms as the linkage between the anthraquinone core and the pendent aliphatic and/or aromatic groups in both the 2,6 and 2,7 positions were found to be surprisingly soluble ($\leq 15\%$) in host nematic mixtures of positive and negative dielectric anisotropy, see Table 3.22. Many of these thio-substituted anthraquinone dyes exhibit high order parameters ($S_{dye} \approx 0.8$) and absorption coefficients in nematic mixtures, as well as exhibiting a full range of colours.[225,247,248] The colours exhibited by anthraquinone dyes cover some part of all of visible spectrum depending on the number, position and nature of the substituents. For example, some compounds with two pendent groups in position 2,6 or 2,7 such as **177** and **178**, exhibit a yellow colour, some dyes with three substituents, such as **179**, are orange and some dyes with four substituents such as **180** are red. Dyes with cyclohexyl substituents are very soluble in various nematic host mixtures, *e.g.* compare compounds **177** and **178**. The high order parameter of these anthraquinone dyes is surprising to some degree. However, the long molecular axis of suitably substituted anthraquinone dyes may still be more-or-less parallel to the director of the host nematic mixture. This appears to be valid for 1,5-disubstituted anthraquinones and 2,6-disubstituted anthraquinones shown in Tables 3.20–3.22. However, the high order parameter of the tri- and tetra-thiosubstituted anthraquinone dyes, such as **179** and **180**, is still surprising.

Positive contrast GH-LCDs with black figures against a white background exhibit a higher contrast than similar GH-LCDs with coloured figures against a white background. However, this requires black dye mixtures.[246–251] This can be achieved by mixing several of the anthraquinone dyes collated in Tables 3.19–3.22, but especially the thio-substituted dyes shown in Table 3.22, to cover the whole of the visible spectrum.[37,38] These dyes have absorption bands between 50–150 nm broad. Therefore, three dyes are sufficient to absorb almost all wavelengths in the visible spectrum (380–780 nm).[246–251]

Negative Contrast Dyes

The tetralins (**181–183**) collated in Table 3.23 are red liquid crystals[252,253] with a transition moment orthogonal to the molecular long axis, see Figure 3.13. Since they are liquid crystalline themselves they are very soluble in nematic host mixtures. Such mixtures exhibit high negative dichroic ratios and can be used in White and Taylor GH-LCDs with positive contrast. Furthermore, mixtures containing an additional anthraquinone dye of positive dichroic ratio can be used to switch from one colour to another under the action of an electric field.

The azulenes (**184** and **185**) also shown in Table 3.23 are red liquid crystalline dyes with a negative dichroic ratio. The compound **184** exhibits a low order parameter, probably due to its strongly non-linear shape.[254] However, the longer, thinner compound **185** results in the formation of an unidentified smectic phase.[255] Both of these classes of azulene dyes are very soluble in nematic host mixtures suitable for GH-LCDs.

Nematic Liquid Crystals — Hosts

Analysis of the electro-optical characteristics of GH-LCDs showed[214] that for all types of guest–host chiral nematic phase change effects the contrast is greatest when

$$\frac{\Delta np}{\lambda} \ll 1 \tag{37}$$

where p is the pitch of the chiral nematic host. However, since p should be as large as possible in order to minimise the value of the voltage required to unwind the helix, then Δn of the chiral nematic host should be as low as possible. This is fortunate since nematic hosts are required with a low UV absorption above the glass cut-off point in order to minimise photochemical degradation of the chromophore. Both physical properties are lower for compounds incorporating aliphatic rings, such as cyclohexane, than for analogous materials with aromatic phenyl rings. In the absence of polarisers and dyes, liquid crystal molecules with an aromatic core absorb UV light and are degraded photochemically. However, in nematic mixtures containing organic dye molecules, energy transfer from the liquid crystal molecules in the excited state to the chromophore in the ground state leads to accelerated photodegradation of the photosensitised dye as well as decomposition of the host nematic mixture. Conversely, the degradation of the nematic host is retarded.[231] This degradation can be minimised by using nematic hosts with no aromatic rings. Furthermore, in order to maximise the contrast between the on-state and the off-state the order parameter of the chiral nematic mixture and dyes should be as high as possible. This implies a high nematic clearing point, see Chapter 2.[237] As always the viscosity should be as low as possible in order to minimise switching times. They should be of high positive dielectric anisotropy for GH-LCDs of negative contrast and of negative dielectric anisotropy for GH-LCDs of positive contrast.

Nematic Liquid Crystal Hosts of Positive Dielectric Anisotropy

The first nematic guest–host prototype nematic GH-LCD reported by Heilmeier and Zanoni[211] contained methyl red (157) as the dichroic dye dissolved in 4-butoxybenzoic acid[256] as the nematic liquid crystal host. Other hosts investigated later[212] included 4-methoxycinnamic acid[257] and 4-ethoxy-4-aminobenzonitrile[76,77] (28), see Table 3.4. The melting point of these three single components is very high. Therefore, prototype GH-LCDs had to be operated and evaluated at very high temperatures. Thermal decomposition of the mixtures led sequentially to lower contrast, homeotropic orientation due to decomposition products and finally device breakdown. However, these initial experiments were sufficient to demonstrate the feasibility of this display type.

The advent of the apparently superior White and Taylor device retarded the development of better guest–host materials and mixtures for this device type.

White and Taylor found it was much more practical to use a room temperature nematic mixture containing MBBA (7), the corresponding ethyl homologue *N*-(ethoxybenzylidene)-4'-butylaniline (EBBA) and compound (28), see Table 3.4, reported earlier for use in TN-LCDs. This mixture was doped with an optically active Schiff's base as chiral dopant and various dichroic dyes.[213] The azo-thiazole dye (158) was found to generate the highest order parameter (0.71 at 25°C) in a related mixture consisting of cyano-substituted Schiff's bases and another optically active Schiff's base as chiral dopant. This mixture exhibits a chiral nematic phase with a short pitch (3 μm) at room temperature due to the helical twisting effect of the optically active material. This mixture has a relatively high threshold voltage (≈ 6 V) in a 12 μm test cell with an internal mirror. The contrast ratio between the blue or blue-green off-state to the white on-state was about 5:1. Brightness was high due to the absence of polarisers and the internal mirror.

Most of the dyes developed for GH-LCDs were evaluated in commercial nematic host mixtures such as E7 and E8 from the British Drug Houses company (now part of E. Merck) and related mixtures from F. Hoffmann-La Roche. Although the birefringence of these mixtures is high ($\Delta n \approx 2.2$) and, therefore, not very suitable for commercial GH-LCDs, they allowed comparisons to be made between the data obtained for different dyes in very similar host mixtures. The merocyanines and a wide variety of related dichroic dyes, see Table 3.19, were evaluated and compared in a simple tertiary mixture of Schiff's bases, although this mixture was of no real commercial relevance, it allowed dyes to be compared in the same matrix and promising dyes identified. This is also true of a nematic mixture composed of 4-cyanophenyl esters and 4-*n*-alkoxyphenyl esters of *trans*-4-*n*-alkylcyclohexane acids.[121] Red guest–host nematic mixtures containing dyes of negative dichroism possess attractive response times in prototype GH-LCDs.

Commercial nematic mixtures of positive dielectric anisotropy now contain nematics of low birefringence such as 50, see Table 3.13. The birefringence of nematic liquid crystals containing cyclohexane rings, such as 41 and 50, in place of aromatic rings is lower than that of the fully aromatic compound such as 36 due to the absence of polarisable π electrons. The dioxane derivative (49) combines a relatively low value of birefringence with a high positive dielectric anisotropy. This promotes a high dye order parameter under the action of an electric field.

Nematic Liquid Crystals Hosts of Negative Dielectric Anisotropy

A binary mixture of 4-pentylphenyl 4-pentylbicyclo[2.2.2]octane and 4-heptyl-phenyl 4-pentylbicyclo[2.2.2]octane,[49,126] see Table 3.8, exhibits a wide nematic phase at room temperature with a high clearing point, a low birefringence (0.08), a moderate viscosity (46 cP) and a weakly negative dielectric anisotropy ($\Delta\varepsilon \approx -1.1$) measured at 20°C.[247] It is a surprisingly good solvent for anthraquinone dyes such as the blue anthraquinone dye (171). A 12 μm cell with the Heilmeier and Zanoni configuration with one polariser and homeotropic alignment containing a mixture of 1 wt% of the dye 171 and the binary mixture and the binary ester mixture exhibits positive contrast (blue figures on a white background), a low threshold voltage (4.5 V), short response times (t_{on} = t_{off} = 0.5 s at 9 V) and acceptable contrast (5:1) at 20°C.

8   In-Plane Switching (IPS) LCDs[258–261]

A novel electro-optical display device based on the response of a nematic liquid crystal to electric fields in the plane of the display was reported by researchers at the Fraunhofer-Institut für Festkörper Physik in Freiburg, Germany in the early 1990s.[258–261] This type of LCD does not exhibit the strong viewing angle dependency and asymmetry of optical contrast observed for ECB-LCDs, TN-LCDs and STN-LCDs, since the optical axis of the nematic liquid crystal lies in the plane of the cell. Therefore, it aroused considerable interest as a possibility to overcome the limitations of commercial LCDs, although field effects utilising interdigitated electrodes had been reported previously.[262,263] There are several variants of IPS-LCDs. Some of these, however, display a pattern of stripe defects at the electrodes due to fringing effects. Therefore, the most promising type utilises a nematic liquid crystal of negative dielectric anisotropy in the original display configuration.[258–261] The electro-optical characteristic allows grey-scale, *i.e.* full colour. In-plane switching (IPS) LCDs utilising this config-uration with active matrix addressing are currently being manufactured, especially as computer monitors due to their wide viewing angles and good contrast.[264–268] The dark state is black with the absence of interference colours even without a compensation (optical retarder) film.

Display Configuration

A nematic liquid crystal mixture is contained in a liquid crystal cell with non-transparent interdigitated finger electrodes on one substrate of the cell as shown in Figure 3.21. The second substrate does not carry an electrode. The narrow stripe electrodes, *e.g.* ≈ 5 μm, are metallic, *e.g.* ITO or chromium, and are separated, *e.g.* 10 μm, to create the active pixel area. Therefore, the aperture ratio is low. Consequently the display brightness may also be low in the absence of strong backlighting. A zig-zag configuration of the electrodes can further suppress interference colours. A homogeneous azimuthal alignment of the nematic liquid crystal director with a non-zero zenithal pretilt angle ($\theta \approx 3$–5°) is obtained using rubbed polyimide at both surfaces. The relatively high pretilt angle is in order to avoid reverse tilt. The nematic director is parallel at both surfaces and throughout the cell. The polarisers are crossed with the first polariser set parallel to the optic axis at the first substrate surface.

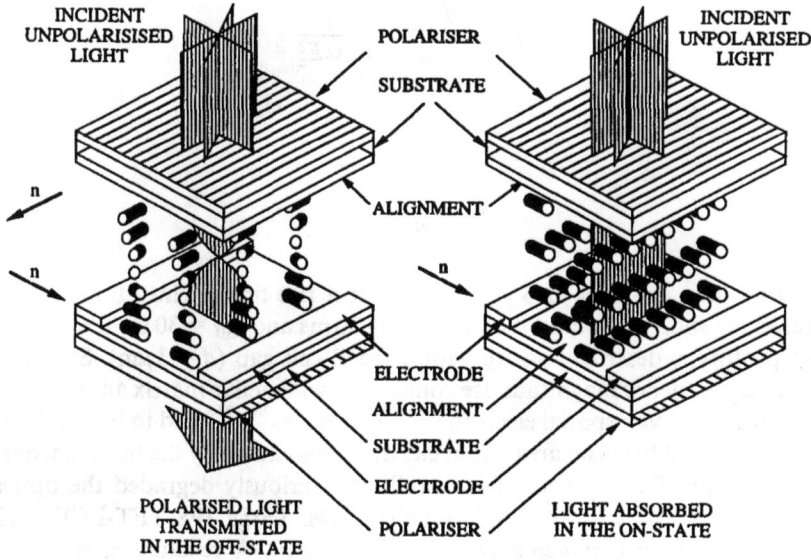

Figure 3.21   *Schematic representation of an in-plane switching liquid crystal display (IPS-LCD).*[258–261]

Off-State

The homogeneously aligned nematic liquid crystal mixture does not rotate the plane polarised light generated by the first polariser. Therefore, the light is absorbed by the analyser. Very efficient absorption of light takes place.

On-State

The application of an electric field between the electrodes causes the nematic liquid crystal mixture to reorient itself with the molecular long axis parallel to the applied field. The strong anchoring at the rubbed polyimide surfaces maintains the original parallel alignment at the surfaces. This results in two areas of twist as shown in Figure 3.21. The direction of twist of the two regions of twist are opposite. However, the upper area of twist dominates the optics of the cell in the on-state as long as the ratio of the pitch of the upper area of twist (P_T) to the lower area of reverse twist (P_R) is greater than 5:1. A twist angle of up to 90° can be produced. The amplitude of transmission in the first minimum ($u = \sqrt{3}$) is then relatively high. However, the influence of the area of reverse twist increases for the other Gooch and Tarry minima (*i.e.* $\sqrt{15}, \sqrt{35}, \sqrt{63}$, *etc.*).

The switch-on time, t_{on}, and the switch-off time, t_{off}, are both proportional to the square of the cell gap, d, as for the TN-LCD,[264]

$$t_{on} = \frac{\gamma_1}{\varepsilon_0 \Delta\varepsilon(E^2 - E_c^2)} \tag{38}$$

$$t_{\mathrm{off}} = \frac{d^2 \gamma_1}{\pi^2 k_{22}} = \frac{\gamma_1}{\varepsilon_0 \Delta \varepsilon E_{\mathrm{c}}^2} \tag{39}$$

where

$$E_{\mathrm{c}} = \frac{\pi}{d} \sqrt{\frac{k_{22}}{\varepsilon_0 \Delta \varepsilon}} \tag{40}$$

Therefore, thin cell gaps also give rise to short rise times. Thus, thin cells are required for short response times, *e.g.* $t_{\mathrm{on}} = 20\,\mathrm{ms}$ and $t_{\mathrm{off}} = 30\,\mathrm{ms}$ for the IPS-LCD prototype described above with a small cell gap ($d = 3\,\mu\mathrm{m}$). Switching times of this order of magnitude are compatible with active matrix addressing at video rate. However, special electrode configurations[264–266] had to be developed in order to shield the pixel areas from electric fields caused by the bus electrodes. These fringing effects generated crosstalk and seriously degraded the optical response of the first prototype IPS-TFT-LCDs. Large IPS-TFT-LCDs (13″ diagonal) are now commercially available with XGA resolution, see Table 3.24.[267,268] The data for a CRT and a TN-TFT-LCD are also collated in Table 3.24 for comparison purposes. In the IPS-TFT-LCDs the polariser orientation is inverted in order to produce black information on a white background as opposed to the white information on a black background described above.

Nematic Materials

Nematic liquid crystals of negative dielectric anisotropy with a low rotational viscosity and a high value of the twist elastic constant k_{22} are required, see Equations 38–40. Furthermore, a high value for the resistivity and short

Table 3.24 *Specifications for a 15″ CRT, a 13.3″ IPS-TFT-LCD with XGA resolution and a 12.1″ TN-TFT-LCD with XGA resolution (1024 × 768 × 3 pixels)*[267,268]

Property	CRT	TN-TFT-LCD	IPS-TFT-LCD
Approximate display weight	13 kg	0.5 kg	1 kg
Display volume	48 858 cm³	468 cm³	1080 cm³
External dimensions (width × height × depth)	36.2 × 35.8 × 37.7 cm	27.6 × 20.0 × 0.9 cm	33.8 × 26.4 × 1.3 cm
Effective display area	270 × 200 mm	246 × 184 mm	270 × 202 mm
Pixel size	0.28 × 0.28 mm	0.26 × 0.26 mm	0.26 × 0.09 mm
Number of pixels	6 888 000 (RGB)	2 359 296 (RGB)	2 359 296 (RGB)
Number of colours	> 260 000	> 260 000	> 260 000
Vertical viewing angle	±90°	+10° and −30°	±70°
Horizontal viewing angle	±90°	±45°	±70°
Brightness	100 cd m⁻²	70 cd m⁻²	120 cd m⁻²
Power consumption	85 W	4 W	18 W
Response time ($t_{\mathrm{on}} + t_{\mathrm{off}}$)		80 ms	70 ms
Contrast ratio	> 100:1	> 100:1	> 100:1

response times, *i.e.* $t_{on} + t_{off} \leq 50$ ms, are necessary for active matrix addressing, *e.g.* on a TFT substrate. Appropriate mixtures are commercially available. They contain difluoro-substituted liquid crystals of negative dielectric anisotropy, see Table 3.3.

9 References

1 G. H. Heilmeier, L. A. Zanoni and L. Barton, *Proc. IEEE*, 1968, **56**, 1162.
2 W. Helfrich, *J. Chem. Phys.*, 1969, **51**, 4092.
3 W. Helfrich, *Mol. Cryst. Liq. Cryst.*, 1973, **21**, 187.
4 G. H. Heilmeier and W. Helfrich, *Mol. Cryst. Liq. Cryst.*, 1970, **16**, 155.
5 G. Durand, M. Veyssié, F. Rondelez and L. Leger, *C. R. Acad. Sci. Paris*, 1970, **270B**, 97.
6 V. Zwetkoff, *Acta Physiochim. USSR*, 1937, **6**, 885.
7 R. Williams, *J. Chem. Phys.*, 1963, **39**, 384.
8 G. Hansen, unpublished dissertation, 1907, Halle, Germany.
9 J. A. Castellano, J. E. Goldmacher, L. A. Barton and J. S. Kane, *J. Org. Chem.*, 1968, **33**, 3501.
10 A. Aviram, R. J. Cox and W. R. Young, *Angew. Chem. Int. Ed. Eng.*, 1971, **10**, 410.
11 J. Billard, J. Jacques, M. Leclerq and J. Malthete, *Comptes Rendus. Acad. Sci. C Paris*, 1971, **273**, 291.
12 A. Aviram, I. Haller and W. R. Young, *Mol. Cryst. Liq. Cryst.*, 1971, **13**, 357.
13 R. Steinsträsser, *Z. Naturforsch.*, 1972, **276**, 774.
14 R. Steinsträsser and L. Pohl, *Z. Naturforsch.*, 1971, **266**, 577.
15 H. Kelker and B. Scheurle, *Angew. Chem. Int. Ed. Eng.*, 1969, **8**, 884.
16 C. Weygand and R. Gabler, *Ber. Dtsch. Chem. Ges.*, 1938, **71**, 2399.
17 J. van der Veen, W. H. de Jeu, A. H. Grobben and J. Boven, *Mol. Cryst. Liq. Cryst.*, 1972, **17**, 291.
18 W. Greubel and U. Wolff, *Appl. Phys. Lett.*, 1971, **19**, 213.
19 D. P. McLemore and E. F. Carr, *J. Chem. Phys.*, 1972, **37**, 3245.
20 E. F. Carr, *Mol. Cryst. Liq. Cryst.*, 1969, **7**, 253.
21 W. H. de Jeu, C. J. Gerritsma and A. M. van Boxtel, *Phys. Lett.*, 1971, **34A**, 203.
22 W. J. A. Goossens, *Phys. Lett.*, 1972, **40A**, 95.
23 W. H. de Jeu and C. J. Gerritsma, *J. Chem. Phys.*, 1972, **56**, 4752.
24 R. B. Meyer, *Appl. Phys. Lett.*, 1968, **20**, 1024.
25 P. G. de Gennes, *Solid State Commun.*, 1968, **6**, 163.
26 R. B. Meyer, *Appl. Phys. Lett.*, 1969, **14**, 208.
27 F. J. Kahn, *Phys. Rev. Lett.*, 1970, **24**, 209.
28 J. J. Wysocki, J. Adams and W. Haas, *Phys. Rev. Lett.*, 1968, **20**, 1024.
29 J. Adams, W. Haas and J. J. Wysocki, *Mol. Cryst. Liq. Cryst.*, 1969, **8**, 9.
30 J. Adams, W. Haas and J. J. Wysocki, *Mol. Cryst. Liq. Cryst.*, 1969, **8**, 471.
31 H. Baessler and M. M. Labes, *Phys. Rev. Lett.*, 1968, **21**, 1791.
32 E. Jakeman and E. P. Raynes, *Phys. Lett.*, 1972, **39A**, 69.
33 P. R. Gerber, *Z. Naturforsch.*, 1982, **38A**, 407.
34 M. F. Schiekel and K. Fahrenschon, *Appl. Phys. Lett.*, 1971, **15**, 391.
35 H. Mailer, K. L. Likins, T. R. Taylor and J. L. Ferguson, *Appl. Phys. Lett.*, 1971, **118**, 105.
36 M. Hareng, G. Assouline and F. Leiba, *Appl. Opt.*, 1972, **11**, 2920.
37 F. J. Kahn, *Appl. Phys. Lett.*, 1972, **20**, 199.
38 G. Labrunie and J. Robert, *J. Appl. Phys.*, 1973, **44**, 4869.
39 H-p. Schad, *SID '82 Digest*, 1982, 244.
40 V. Fréedericksz and V. Zolina, *Trans. Faraday Soc.*, 1933, **29**, 919.
41 W. Helfrich, *Mol. Cryst. Liq. Cryst.*, 1973, **21**, 187.

42 L. M. Blinov, in 'Handbook of Liquid Crystal Research', Eds. P. J. Collings and J. S. Patel, Oxford University Press, Oxford, 1997, p. 125.

43 S. T. Wu, *Appl. Phys. Lett.*, 1990, **57**, 986.

44 S. T. Wu, *Mol. Cryst. Liq. Cryst.*, 1991, **207**, 1.

45 A. Rapini and M. Papoular, *J. de Phys. (Paris)*, 1969, **30**, C4–54.

46 F. Gharadjedaghi, *Disp. Technol.*, 1986, **1**, 95.

47 R. Steinsträsser and F. Del Pino, *Ger. Pat.*, DE-24 500 088, 1974.

48 T. Inukai, H. Inone, K. Furukawa, S. Saito, S. Sugimori and K. Yokohama, DE-2 937 700, 1979.

49 G. W. Gray and S. M. Kelly, *Mol. Cryst. Liq. Cryst.*, 1981, **75**, 109.

50 M. A. Osman, *Mol. Cryst. Liq. Cryst.*, 1982, **82**, 295.

51 S. M. Kelly and H-p. Schad, *Mol. Cryst. Liq. Cryst.*, 1984, **110**, 239.

52 S. M. Kelly and T. Huynh-Ba, *Helv. Chim. Acta*, 1983, **66**, 1850.

53 M. A. Osman and T. Huynh-Ba, *Mol. Cryst. Liq. Cryst. Lett.*, 1983, **92**, 57.

54 S. M. Kelly and H-p. Schad, *Helv. Chim. Acta*, 1985, **68**, 813.

55 Hp. Schad and S. M. Kelly, *Mol. Cryst. Liq. Cryst.*, 1986, **75**, 133.

56 M. A. Osman and T. Huynh-Ba, *Mol. Cryst. Liq. Cryst. Lett.*, 1983, **82**, 339.

57 M. A. Osman, *Mol. Cryst. Liq. Cryst.*, 1985, **128**, 45.

58 M. A. Osman, *Helv. Chim. Acta*, 1985, **68**, 606.

59 M. Schadt, *Proc. Japan Display '83, Kobe*, 1983, 220.

60 M. Schadt, M. Petrzilka, P. R. Gerber, A. Villiger and G. Trickes, *Mol. Cryst. Liq. Cryst.*, 1983, **94**, 139.

61 M. P. Burrow, G. W. Gray, D. Lacey and K. J. Toyne, *Z. Chem.*, 1986, **26**, 21.

62 R. Eidenschink, O. Haas, M. Römer and B. S. Scheuble, *Angew. Chem.*, 1984, **96**, 151.

63 R. Eidenschink and B. S. Scheuble, *Mol. Cryst. Liq. Cryst. Lett.*, 1986, **3**, 33.

64 V. Reiffenrath, J. Krause, H. J. Plach and G. Weber, *Liq. Cryst.*, 1989, **5**, 159.

65 V. Reiffenrath, and M. Bremer, *Angew. Chem. Int. Ed. Eng.*, 1994, **33**, 87.

66 S. M. Kelly, *Liq. Cryst.*, 1991, **10**, 261.

67 M. Schadt and W. Helfrich, *Appl. Phys. Lett.*, 1971, **18**, 127.

68 Ch. Maugin, *Bull. Soc. Fr. Minéral. Cristallogr. Phys. Z.*, 1911, **12**, 1011.

69 C. H. Gooch and H. A. Tarry, *Electron. Lett.*, 1974, **10**, 2.

70 L. Pohl, G. Weber, R. Eidenschink, G. Baur and W. Fehrenbach, *Appl. Phys. Lett.*, 1981, **38**, 497.

71 J. van der Veen, W. H. de Jeu, A. H. Grobben and J. Boven, *Mol. Cryst. Liq. Cryst.*, 1972, **17**, 291.

72 E. Jakeman and E. P. Raynes, *Phys. Lett.*, 1972, **39A**, 69.

73 F. M. Jaeger, *Recueil Trav. Chim. Pays-Bas*, 1907, **26**, 311.

74 G. W. Gray and D. G. McDonnell, *Mol. Cryst. Liq. Cryst. Lett.*, 1977, **34**, 211.

75 G. W. Gray and D. G. McDonnell, *Mol. Cryst. Liq. Cryst.*, 1976, **37**, 189.

76 E. Froehlich, Dissertation, Halle, 1910.

77 G. H. Heilmeier, L. A. Zanoni and J. E. Goldmacher, in 'Liquid Crystals and Ordered Fluids', Eds. J. F. Johnson and R. S. Porter, Plenum Press, New York, 1970.

78 J. A. Castellano, *Ger. Pat. Appl.*, DE-OS 1 928 003, 1971.

79 L. Pohl and R. Steinsträsser, *Ger. Pat. Appl.*, DE-OS 2 024 269, 1971.

80 R. Steinsträsser, *Z. Naturforsch.*, 1972, **27b**, 774.

81 A. Boller, H. Scherrer M. Schadt and P. Wild, *Proc. IEEE*, 1972, **60**, 1002.

82 G. W. Gray, K. J. Harrison and J. A. Nash, *Electron. Lett.*, 1973, **9**, 130.

83 A. I. Pavluchenko, V. V. Titov and N. I. Smirnova, in 'Advances in Liquid Crystal Research and Applications', Ed. L. Bata, Pergamon Press, Oxford, 1980, 1007.

84 L. A. Karamysheva, E. I. Kovshev, A. I. Pavluchenko, K. V. Roitman, V. V. Titov, S. I. Torgova and M. F. Grebyonkin, *Mol. Cryst. Liq. Cryst.*, 1981, **67**, 241.

85 A. Boller. M. Cereghetti, M. Schadt and H. Scherrer, *Mol. Cryst. Liq. Cryst.*, 1977, **42**, 1225.

86 R. Eidenschink, D. Erdmann, J. Krause and L. Pohl, *Angew. Chem.*, 1977, **89**, 103.
87 A. I. Pavluchenko, N. I. Smirnova, V. F. Petrov, M. F. Grebyonkin and V. V. Titov, *Mol. Cryst. Liq. Cryst.*, 1991, **209**, 155.
88 M. F. Grenyonkin, V. F. Petrov and B. I. Ostrovsky, *Liq. Cryst.*, 1990, **7**, 367.
89 V. Petrov, S. I. Torgova, L. A. Karamysheva and S. Tanenaka, *Liq. Cryst.*, 1999, **26**, 1141.
90 J. Bartulin, C. Zuniga, H. Muller, T. R. Taylor and W. Haase, *Mol. Cryst. Liq. Cryst.*, 1987, **150B**, 237.
91 T. Inukai, H. Inoue and H. Sato, *USA Patent*, 4 211 666, 1978.
92 M. A. Osman and L. Revesz, *Mol. Cryst. Liq. Cryst.*, 1982, **82**, 41.
93 H. Sorkin, *Mol. Cryst. Liq. Cryst.*, 1980, **56**, 279.
94 V. S. Bezborodov, *Zh. Org. Chim.*, 1989, **25**, 2168.
95 Y. Haramoto and H. Kamogawa, *Chem. Lett.*, 1985, 79.
96 G. W. Gray and S. M. Kelly, *Angew. Chem. Int. Ed.*, 1981, **20**, 393.
97 A. Villiger, A. Boller and M. Schadt, *Z. Naturforsch.*, 1979, **34B**, 1535.
98 G. W. Gray, *Abstract, 12th Arbeitstagung Flüssigkristalle, Freiburg, Germany*, 1978.
99 R. Eidenschink, D. Erdmann, J. Krause and L. Pohl, *Angew. Chem.*, 1978, **90**, 133.
100 W. Sucrow and W. Schatull, *Z. Naturforsch.*, 1982, **37b**, 1336.
101 M. Petrzilka and M. Schadt, *US Patent*, 4 565 425, 1986.
102 H.-J. Deutscher, F. Kuschel, S. König, H. Kresse, D. Pfeiffer, A. Wiegeleben, J. Wulf and D. Demus, *Z. Chem.*, 1977, **17**, 64.
103 S. Sugimori, H. Sato, T. Inukai and K. Furakawa, *Ger. Pat. Appl.*, DE-OS 3 029 378, 1981.
104 N. Carr, G. W. Gray and D. G. McDonnell, *Mol. Cryst. Liq. Cryst.*, 1983, **97**, 13.
105 M. Petrzilka, *Mol. Cryst. Liq. Cryst.*, 1984, **111**, 329.
106 S. M. Kelly, *Liq. Cryst.*, 1991, **10**, 273.
107 R. Buchecker and M. Schadt, *Mol. Cryst. Liq. Cryst.*, 1987, **149**, 359.
108 M. Petrzilka and M. Schadt, *Mol. Cryst. Liq. Cryst.*, 1985, **131**, 109.
109 S. M. Kelly, A. Germann, R. Buchecker and M. Schadt, *Liq. Cryst.*, 1994, **16**, 67.
110 S. M. Kelly, M. Schadt and H. Seiberle, *Liq. Cryst.*, 1995, **18**, 581.
111 M. Schadt, R. Buchecker, A. Villiger, F. Leenhouts and J. Fromm, *Trans. IEEE*, 1986, **ED-33**, 1187.
112 M. Schadt, M. Petrzilka, P. R. Gerber and A. Villiger, *Mol. Cryst. Liq. Cryst.*, 1985, **122**, 241.
113 M. Schadt, R. Buckecker and K. Müller, *Liq. Cryst.*, 1989, **5**, 293.
114 G. W. Gray and S. M. Kelly, *J. Mater. Chem.*, 1999, **9**, 2037.
115 A. J. Leadbetter, R. M. Richardson and C. N. Collings, *J. de Phys.*, 1975, **36**, 37.
116 Hp. Schad and M. A. Osman, *J. Chem. Phys.*, 1981, **75**, 880; 1983, **79**, 5710.
117 Hp. Schad and S. M. Kelly, *J. Chem. Phys.*, 1984, **81**, 1514.
118 C. J. F. Boettcher and P. Bordwijk, in 'Theory of Electric Polarisation', Elsevier, Amsterdam, Vol. II, 1978.
119 P. Bordwijk, *J. Chem. Phys.*, 1980, **73**, 595.
120 M. J. Bradshaw and E. P. Raynes, *Mol. Cryst. Liq. Cryst.*, 1983, **91**, 145.
121 H.-J. Deutscher, B. Laaser, W. Dölling and H. Schubert, *J. Prakt. Chem.*, 1978, **320**, 194.
122 M. E. Neubert, L. T. Carlino, D. L. Fishel and R. M. D'Sidocky, *Mol. Cryst. Liq. Cryst.*, 1980, **59**, 253.
123 G. W. Gray, C. Hogg and D. Lacey, *Mol. Cryst. Liq. Cryst.*, 1981, **67**, 1.
124 M. A. Osman and L. Revesz, *Mol. Cryst. Liq. Cryst.*, 1980, **56**, 157.
125 G. W. Gray and S. M. Kelly, *Mol. Cryst. Liq. Cryst.*, 1981, **75**, 95.
126 N. Carr, G. W. Gray and S. M. Kelly, *Mol. Cryst. Liq. Cryst. Lett.*, 1985, **1**, 53.
127 M. A. Osman, *Mol. Cryst. Liq. Cryst.*, 1985, **128**, 45.
128 R. Eidenschink, *Kontakte (Darmstadt)*, 1979, **1**, 15.
129 R. Eidenschink, M. Römer and F. V. Allen, in 'Liquid Crystals and Ordered Fluids', Eds. A. C. Griffin and J. F. Johnson, Plenum Press, New York, 1984, p. 737.

130 M. A. Osman, *Z. Naturforsch.*, 1983, **38a**, 693.

131 M. A. Osman, *Mol. Cryst. Liq. Cryst.*, 1982, **72**, 291.

132 K. Praefcke, D. Schmidt and G. Heppke, *Chem. Zeit.*, 1980, **104**, 269; *Is. J. Chem.*, 1979, **18**, 195.

133 R. Eidenschink, *Mol. Cryst. Liq. Cryst.*, 1985, **113**, 57.

134 R. Eidenschink, M. Römer, G. Weber, G. W. Gray and K. J. Toyne, *German Patent*, DE 3321373, 1983.

135 N. Carr and G. W. Gray, *Mol. Cryst. Liq. Cryst.*, 1985, **124**, 27.

136 H. Takatsu, K. Takeuchi and H. Sato, *Japan Display '83*, 1983, 228.

137 H.-J. Deutscher, M. Körber and H. Schubert, 'Advances in Liquid Crystal Research and Application', Ed. L. Bata, Pergamon Press, Oxford, 1980, p. 1075.

138 E. Poetsch, V. Reiffenrath and B. Rieger, *24ᵗʰ Arbeitstagung Flüssigkristalle, Freiburg, Germany*, 1995.

139 B. S. Scheuble, *Proc. 16ᵗʰ Freiburger Arbeitstagung Flüssigkristalle, Freiburg, Germany*, 1987.

140 H. Takatsu, K. Takeuchi and H. Sato, *Mol. Cryst. Liq. Cryst.*, 1983, **100**, 345.

141 G. W. Gray and D. G. McDonnell, *Mol. Cryst. Liq. Cryst.*, 1979, **53**, 147.

142 R. Eidenschink and M. Römer, *Proc. 13ᵗʰ Freiburger Arbeitstagung Flüssigkristalle, Freiburg, Germany*, 1983.

143 U. Finkenzeller, A. Kurmeier and E. Poetsch, *Proc. 18ᵗʰ Freiburger Arbeitstagung Flüssigkristalle, Freiburg, Germany*, 1989.

144 E. Bartmann, D. Dorsch and U. Finkenzeller, U. Kurmeier and E. Poetsch, *Proc. 19ᵗʰ Freiburger Arbeitstagung Flüssigkristalle, Freiburg, Germany*, 1990.

145 B. Rieger, E. Poetsch and V. Reifenrath, *Proc. 19ᵗʰ Freiburger Arbeitstagung Flüssigkristalle, Freiburg, Germany*, 1990.

146 A. I. Pavluchenko, N. I. Smirnova, V. F. Petrov, Y. A. Fialkov, S. V. Shelyazhenko and L. M. Yagupolsky, *Mol. Cryst. Liq. Cryst.*, 1991, **209**, 225.

147 E. Bartmann, D. Dorsch and U. Finkenzeller, *Mol. Cryst. Liq. Cryst.*, 1991, **204**, 77.

148 Y. Goto, T. Ogawa, S. Sawada and S. Sugimori, *Mol. Cryst. Liq. Cryst.*, 1991, **209**, 1.

149 Y. Onji, M. Ushioda, S. Matsui, T. Kondo and Y. Goto, *Eur. Pat. Appl.*, EPA 0 647 696 A1, 1993.

150 K. Tarumi, M. Bremer and T. Geelhaar, *Ann. Rev. Sci.*, 1997, **27**, 423.

151 H. Takatsu, K. Takeuchi, M. Sasaki and H. Onishi, *Mol. Cryst. Liq. Cryst.*, 1991, **206**, 159.

152 E. Bartmann, J. Krause and K. Tarumi, *Proc. 23ʳᵈ Freiburger Arbeitstagung Flüssigkristalle, Freiburg, Germany*, 1994.

153 M. Bremer, *Adv. Mater.*, 1995, **7**, 867.

154 R. Buchecker, G. Marck and M. Schadt, *Mol. Cryst. Liq. Cryst.*, 1995, **260**, 93.

155 E. Bartmann, *Adv. Mater.*, 1996, **8**, 570.

156 M. Goulding, S. Greenfield, O. Parri and D. Coates, *Mol. Cryst. Liq. Cryst.*, 1995, **265**, 27.

157 S. Greenfield, D. Coates, M. Goulding and R. Clemitson, *Liq. Cryst.*, 1995, **18**, 665.

158 D. W. Berreman and W. R. Heffner, *J. Appl. Phys.*, 1985, **122**, 1.

159 D. W. Berreman and W. R. Heffner, *J. Appl. Phys.*, 1981, **52**, 3032.

160 R. N. Thurston, *Mol. Cryst. Liq. Cryst.*, 1985, **122**, 1.

161 C. M. Waters, E. P. Raynes and V. Brimmell, *Japan Display '83*, 1983, 396.

162 C. M. Waters, E. P. Raynes and V. Brimmell, *Proc. SID '84*, 1984, **25**, 261.

163 C. M. Waters, E. P. Raynes and V. Brimmell, *Mol. Cryst. Liq. Cryst.*, 1985, **123**, 303.

164 E. P. Raynes, *Mol. Cryst. Liq. Cryst. Lett.*, 1986, **4**, 1.

165 E. P. Raynes and C. M. Waters, *Displays*, 1987, 59.

166 E. P. Raynes, in 'The Optics of Thermotropic Liquid Crystals', Eds. S. Elston and R. Sambles, Taylor and Francis, London, 1998, p. 289.

167 T. J. Scheffer and J. Nehring, *Appl. Phys. Lett.*, 1984, **45**, 1021.

168 T. J. Scheffer, J. Nehring, M. Kaufmann, H. Amstutz, D. Heimgartner and P. Eglin, *SID '85 Digest*, 1985, 120.
169 T. J. Scheffer and J. Nehring, *Ann. Rev. Sci.*, 1997, **27**, 555.
170 N. Kando, T. Nakagomi and S. Hasagawa, *Ger. Pat. Appl.*, DE 3403259, 1985.
171 M. Schadt and F. Leenhouts, *J. Appl. Phys.*, 1987, **50**, 236.
172 F. Leenhouts and M. Schadt, *Mol. Cryst. Liq. Cryst.*, 1988, **158B**, 241.
173 F. Moia, M. Schadt and H. Seiberle, *Proc. SID '93*, 1993, 1.
174 P. M. Alt and P. Pleshko, *IEEE Trans. Electron. Dev.*, 1974, **21**, 146.
175 T. J. Scheffer, B. Clifton, D. Prince and A. R. Connor, *Displays*, 1993, **14**, 74.
176 V. G. Chigrinov, V. V. Belyaev, S. V. Belyaev and M. F. Grebenkin, *Sov. Phys. JETP*, 1979, **50**, 994.
177 A. Göbl-Wunsch, G. Heppke and F. Oestereicher, *J. Phys. (Paris)*, 1979, **40**, 773.
178 S. Naemura, T. Oyama, H. Plach, G. Weber and B. Scheuble, *E. Merck, Liquid Crystal Seminar Data*.
179 D. W. Berreman, *Appl. Phys. Lett.*, 1974, **25**, 12.
180 N. Kimura, T. Shinomiya, K. Yamamoto, Y. Ichimura, K. Nakegawa, Y. Ishi and M. Matsuura, *Proc. SID '88*, 1988, 49.
181 H. Watanabe, O. Okumura, H. Wada, A. Ito, M. Yazaki, M. Nagata, H. Takeshita and S. Morozumi, *Proc. SID '88*, 1988, 416.
182 P. J. Bos, *Proc. SID '94*, 1994, 118.
183 M. Schadt and P. R. Gerber, *Z. Naturforsch.*, 1982, **37a**, 165.
184 W. H. de Jeu and W. A. P. Claasen, *J. Chem. Phys.*, 1977, **67**, 3705.
185 Hp. Schad, G. Baur and G. Meier, *J. Chem. Phys.*, 1979, **70**, 2770.
186 M. Schadt, *Liq. Cryst.*, 1993, **14**, 73.
187 V. Petrov, S. I. Torgova, L. A. Karamysheva and S. Tanenaka, *Liq. Cryst.*, 1999, **26**, 1141.
188 M. Schadt and F. Müller, *Rev. Phys. Appliquée*, 1979, **14**, 265.
189 J. Constant and E. P. Raynes, *Mol. Cryst. Liq. Cryst.*, 1980, **62**, 115.
190 M. Schadt and P. R. Gerber, *Z. Naturforsch.*, 1982, **37a**, 179.
191 J. Constant and E. P. Raynes, *Mol. Cryst. Liq. Cryst.*, 1980, **62**, 115.
192 M. Schadt and M. Petrzilka, *Japan Display '84*, 1984, 53.
193 R. G. Priest, *Phys. Rev.*, 1973, **A7**, 720.
194 J. P. Straley, *Phys. Rev.*, 1973, **A8**, 2181.
195 M. Petrzilka, *Mol. Cryst. Liq. Cryst.*, 1985, **131**, 109.
196 H. Ohnishi, H. Takatsu, K. Takeuchi, F. Moia and M. Schadt, *SID '92 Digest*, 1982, 17.
197 K. Praefcke, D. Schmidt and G. Heppke, *Chem. Zeit.*, 1980, **104**, 269.
198 E. Poetsch, V. Reiffenrath and B. Rieger, *24th Arbeitstagung Flüssigkristalle, Freiburg, Germany*, 1995, P01.
199 M. Petrzilka, R. Buckecker, S. Lee-Schmiederer, M. Schadt and A. Germann, *Mol. Cryst. Liq. Cryst.*, 1987, **148**,123.
200 M. Petrzilka, *USA Patent*, 4676604, 1987.
201 M. Schadt, R. Buckecker and A. Villiger, *Liq. Cryst.*, 1990, **79**, 519.
202 S. M. Kelly, A. Germann and M. Schadt, *Liq. Cryst.*, 1994, **16**, 491.
203 S. M. Kelly, *Liq. Cryst.*, 1994, **17**, 211.
204 R. Eidenschink, G. W. Gray, K. J. Toyne and A. E. F. Waechtler, *Mol. Cryst. Liq. Cryst. Lett.*, 1988, **5**, 177.
205 S. Kawakami, K. Kotani, S. Shirokura, H. Ohnishi, Y. Fujita, T. Ohtsuka, H. Takatsu and F. Moia, *Proc. SID '92, Japan Display*, 1992, 487.
206 H. Seiberle, F. Moia,, M. Schadt, H. Takatsu and K. Kotani, *20th Freiburger Arbeitstagung Flüssigkristalle, Freiburg, Germany*, 1992.
207 L. M. Blinov and V. G. Chigrinov, in 'Electro-optic Effects in Liquid Crystal Materials', Springer, New York, 1994.
208 M. Schadt, *Ann. Rev. Mater. Sci.*, 1997, **27**, 305.
209 D. Coates, E. Merck, private communication.

210 D. Dunmur, in 'Physical Properties of Liquid Crystals', Eds. D. Demus, J. W. Goodby, G. W. Gray, H.-W. Spiess and V. Vill, Wiley-VCH Verlagsgesellschaft GmbH, Weinheim, Germany, 1999.

211 G. H. Heilmeier and L. A. Zanoni, *Appl. Phys. Lett.*, 1968, **13**, 91.

212 G. H. Heilmeier, J. A. Castellano and L. A. Zanoni, *Mol. Cryst. Liq. Cryst.*, 1969, **8**, 293.

213 D. L. White and G. N. Taylor, *J. Appl. Phys.*, 1974, **45**, 4718.

214 T. J. Scheffer and J. Nehring, in 'The Physics and Chemistry of Liquid Crystal Devices', Ed. G. J. Sprokel, Plenum Publishing, New York, 1980, p. 173.

215 T. Uchida, H. Seki, C. Shishido and M. Wada, *IEEE, Trans. Elec. Dev.*, 1979, **ED-26**, 1373.

216 T. Uchida, H. Seki, C. Shishido and M. Wada, *Mol. Cryst. Liq. Cryst.*, 1979, **54**, 161.

217 D. Demus, B. Krücke, F. Kuschel, H. U. Nathninck, G. Pelzl and H. Zaschke, *Mol. Cryst. Liq. Cryst.*, 1979, **56**, 115.

218 P. G. de Gennes, *Solid State Commun.*, 1968, **6**, 163.

219 M. Morita, S. Imamura and J. K. Yatabe, *J. Appl. Phys.*, 1975, **14**, 315.

220 F. Gharadjedaghi, *Mol. Cryst. Liq. Cryst.*, 1981, **68**, 127.

221 F. Fischer, *Z. Naturforsch.*, 1976, **31a**, 41.

222 T. J. Scheffer, in 'Nonemissive Electro-optic Displays', Eds. A. R. Kmetz and F. K. von Willisen, Plenum Press, New York, 1976, p. 45.

223 F. Gharadjedaghi, *J. Appl. Phys.*, 1983, **54**, 4989.

224 I. C. Sage, in 'Handbook of Liquid Crystals', Eds. D. Demus, J. W. Goodby, G. W. Gray, H.-W. Spiess and V. Vill, Wiley-VCH Verlagsgesellschaft GmbH, Weinheim, Germany, 1998, Vol. 1, p. 731.

225 G. W. Gray, *Dyes Pigments*, 1982, **3**, 203.

226 A. Bloom and P. L. K. Hung, *Mol. Cryst. Liq. Cryst.*, 1977, **40**, 213.

227 R. J. Cox, *Mol. Cryst. Liq. Cryst.*, 1979, **55**, 1.

228 M. A. Osman, T. J. Scheffer and H. R. Zeller, *ntz Archiv.*, 1979, **8**, 185.

229 A. C. Lowe, *Mol. Cryst. Liq. Cryst.*, 1980, **59**, 63.

230 L. M. Blinov, V. A. Kizel, V. G. Rumyantsev and V. V. Titov, *J. de Phys., Coll. C1*, 1975, **36**, 69.

231 F. Jones and T. J. Reeve, *Mol. Cryst. Liq. Cryst.*, 1980, **60**, 99.

232 A. C. Lowe and R. J. Cox, *Mol. Cryst. Liq. Cryst.*, 1980, **59**, 63.

233 F. Jones and T. J. Reeve, *Mol. Cryst. Liq. Cryst.*, 1980, **60**, 99.

234 J. Cognard and T. Hieu Phan, *Mol. Cryst. Liq. Cryst.*, 1981, **68**, 207.

235 T. Uchida, C. Shishido, H. Seki and M. Wada, *Mol. Cryst. Liq. Cryst. Lett.*, 1977, **34**, 153; *Mol. Cryst. Liq. Cryst.*, 1977, **39**, 39.

236 R. Eidenschink, *Kontakte (Darmstadt)*, 1984, **2**, 25.

237 W. Maier and A. Saupe, *Z. Naturforsch.*, 1958, **13a**, 564; 1959, **14a**, 882; 1960, **15a**, 287.

238 D. Vorländer, *Trans. Faraday Soc.*, 1933, **29**, 907.

239 J. Constant, E. P. Raynes, I. A. Schanks, D. Coates, G. W. Gray and D. G. McDonnell, *J. Phys. D: Appl. Phys.*, 1978, **11**, 479.

240 J. Constant, M. G. Pellatt and I. H. C. Roe, *Mol. Cryst. Liq. Cryst.*, 1980, **59**, 299.

241 M. Schadt and P. Gerber, *Mol. Cryst. Liq. Cryst.*, 1981, **65**, 241.

242 J. Cognard and T. Hieu Phan, *Mol. Cryst. Liq. Cryst.*, 1981, **70**, 1.

243 S. Aftergut and H. S. Cole, *Appl. Phys. Lett.*, 1981, **38**, 599.

244 G. Heppke, B. Kippenberg, A. Möller and G. Scherowsky, *Proc. Euro Display '81*, 1981, 25.

245 N. Bastürk, J. Cognard and T. Hieu Phan, *Mol. Cryst. Liq. Cryst.*, 1983, **95**, 71.

246 G. Heppke, B. Kippenberg, A. Möller and G. Scherowsky, *Mol. Cryst. Liq. Cryst.*, 1983, **94**, 191.

247 F. C. Saunders, K. J. Harrison, E. P. Raynes and D. J. Thompson, *Proc. SID '82*, 1982, 121.

248 F. C. Saunders, D. J. Thompson K. J. Harrison and E. P. Raynes, *RSRE News Lett. Res. Rev.*, 1982, **44**, 1.
249 J. Cognard ,T. Hieu Phan and N. Bastürk, *Mol. Cryst. Liq. Cryst.*, 1983, **91**, 327.
250 T. J. Scheffer, *J. Appl. Phys.*, 1982, **53**, 257.
251 B. S. Scheuble, G. Weber, L. Pohl and R. E. Jubb, *SID '83 Digest*, 1983, 176.
252 D. Demus, B. Krücke, F. Kuschel, H. U. Nothnick, G. Pelzl and H. Zaschke, *Mol. Cryst. Liq. Cryst. Lett.*, 1979, **56**, 115; D. Demus and H. Zaschke, *Mol. Cryst. Liq. Cryst.*, 1981, **63**, 129.
253 A. Isenberg, B. Krücke, G. Pelzl, H. Zaschke and D. Demus, *Cryst. Res. Technol.*, 1983, **18**, 1059.
254 K. Praefcke and D. Schmidt, *Z. Naturforsch.*, 1981, **36b**, 375.
255 H. Enzenberg, D. Schädler and H. Zaschke, *Z. Chem.*, 1982, **22**, 59.
256 A. E. Bradfield and B. Jones, *J. Chem. Soc. [London]*, 1929, 2660.
257 G. W. Gray and B. Jones, *J. Chem. Soc. [London]*, 1959, 1467.
258 R. Kiefer, B. Weber, F. Windscheid and G. Baur, *Japan Display '92*, 1992, 547.
259 G. Baur, *Proc. 22nd Freiburger Arbeistagung Flüssigkristalle, Freiburg, Germany*, 1993.
260 G. Baur, Proc. *23rd Freiburger Arbeistagung Flüssigkristalle, Freiburg, Germany*, 1994.
261 G. Baur, M. Kamm, R. Kiefer, H. Klausmann, B. Weber, B. Wieber and F. Windscheid, *23rd Freiburger. Arbeistagung Flüssigkristalle, Freiburg, Germany*, 1994.
262 R. A. Soref, *J. Appl. Phys.*, 1974, **45**, 5466.
263 S. Kobayashi, *SID Digest*, 1972, 68.
264 M. O-he, M. Ohta and K. Kondo, *Asia Display '95*, 1995, 707.
265 M. Ohta, K. Kondo and M. O-he, *IEICE Trans. Electron.*, 1996, E79-C.
266 M. Ohta, M. Yoneya and K. Kondo, *EuroDisplay '96*, 1996, 49.
267 M. Ohta, M. Yoneya and K. Kondo, *Proc. SID '96, San Diego*, 1996.
268 G. Baur, *Proc. SID-MID Eur. Meeting, Darmstadt, Germany*, 1996.

CHAPTER 4

Photoluminescence and Electroluminescence from Organic Materials

1 Introduction

Light-emitting diodes (LEDs) are flat-panel display devices, which emit light under the action of an electric current passing through an electroluminescent emissive layer. Electroluminescence in inorganic semiconductors was discovered before the corresponding effect in organic materials was found.[1,2] Consequently the first commercial alpha numeric display devices using electroluminescence used inorganic semiconductor materials, such as GaAsP, in the early 1960s. Simple monochrome inorganic semiconductor LEDs with low information-content are manufactured on a large scale and are found in many electronic instruments, see Chapter 7.

Each pixel is driven directly with two electrodes or by active matrix addressing. This renders high-information-content displays, such as portable computers with up to 3 000 000 pixels, either impractical or very expensive to realise. Even LEDs with a relatively low-information-content are difficult to fabricate cheaply due to the requirement for individual dedicated pixel electrodes. Furthermore, full colour displays would require arrays of alternating pixels of the three primary colours. These would have to be fabricated individually. There remains, therefore, an enormous potential for light-emitting diodes using organic materials due to their advantageous combination of physical properties, above all their ease of processing for displays of almost any size and an almost infinite possibility for modification of the electro-optical properties by suitable design and chemical synthesis. Furthermore, high-information-content LEDs using organic materials can be addressed using either direct addressing, multiplex addressing or active matrix addressing as for LCDs, see Chapter 2.

Electroluminescence from organic materials was first discovered using solid anthracene crystals immersed in a liquid electrolyte.[3–8] Very high voltages, *e.g.* 400–2000 V, were necessary in order to observe electroluminescence partly due to the thickness of the crystals used (≈ 5 mm). Considerably lower operating voltages were found using solid electrodes.[9] However, the decisive break-through was the use of thin, vapour-deposited films and solid electrodes.[10,11] Unfortunately, the efficiency of electroluminescent devices using one layer of light-emitting material sandwiched between two electrodes was very low. Consequently there was little activity in the area of organic electroluminescence for many years until the report by Tang and Van Slyke of the Eastman Kodak Corporation of two-layer devices with a much higher efficiency and lower operating voltage.[12,13] Even higher efficiencies were found for three-layer devices.[14-22] However, despite development over three decades organic light-emitting diodes (OLEDs) are only set now to make a decisive commercial breakthrough.[20] The critical factors inhibiting mass production of OLEDs using small molecules have been too short a display lifetime and insufficient reliability, rather than performance.

Conjugated organic polymers have been prepared and their physical proper-ties studied for several decades. They may exhibit a range of physical properties, such as high electrical conductivity[21,22] and photoconductivity,[23] charge storage,[24] photoluminescence[25] and nonlinear optical response,[26] of interest for a variety of electronic and optoelectronic applications, such as semiconduc-tors, sensors and telecommunications.[27] They have been studied intensively due to their ability to combine the optoelectronic properties of inorganic semicon-ductors with some typical properties of organic polymers, such as ease of processing and patterning, mechanical flexibility and, to some extent, rugged-ness. However, it is the property of electroluminescence that is of the highest interest at the present time.[28-35] Electroluminescent organic polymers possess a number of physical properties compared to inorganic or low-molar-mass electroluminescent materials. Electroluminescent inorganic semiconductors and low-molar-mass organic electroluminescent materials have to be deposited by sublimation or vapour deposition under high vacuum. This is an expensive process and not ideally suited to the mass production of large-area displays. In contrast, uniform thin films of conjugated organic polymers can be deposited from dilute solutions in suitable solvents by the process of spin coating or Doctor blade deposition. Organic polymers form stable glasses, which freeze in the desired structural order without the formation of defects at grain boundaries formed in the process of crystallisation. Furthermore, the production of copolymers from a mixture of different monomers allows the fine-tuning of the physical properties of interest.

The first reports of electroluminescence from an organic polymer failed to attract much attention.[36,37] However, the report of electroluminescence from poly(p-phenylenevinylene) (PPV) by researchers from the Cavendish laboratory in Cambridge, UK,[38,39] and later from a related derivative by researchers from Santa Barbara, USA,[40,41] stimulated a great deal of interest and activity in OLEDs using light-emitting polymers (LEPs). The semiconductor properties of organic conjugated polymers stem from the formation of a π bonding molecular orbital, which corresponds to the highest occupied molecular orbital (HOMO), and a π^* antibonding molecular orbital, which is equivalent to the lowest unoccupied molecular orbital (LUMO). These energy levels are often referred to as a π valence band and a π^* conduction band in analogy with the nomenclature used to describe inorganic semiconductors, such as silicon. The overlap of the *p*-atomic orbitals of sp^2 hybridised atoms, such as carbon or nitrogen, making up the conjugated polymer backbone, gives rise to a substantial degree of delocalisation of electron density. The alternating configuration of conjugated single and double bonds in conjugated polymers gives rise to their semiconductor properties, but also results in a rigid, linear and planar molecular conformation. This tends to give a very high melting point and a very limited solubility in all organic solvents.

Low-molar-mass materials and oligomers with a high degree of conjugation and a propensity to self-assembly are of increasing interest[28] for such applications due to their monodisperse character, flexibility of synthesis, good processability as thin solid films, *etc*. An additional advantage is the improved properties observed due to the propensity of certain classes of these organic materials to self-organise in a supramolecular structure on the desired substrate surface. A serious drawback associated with some low-molar-mass materials and to a lesser extent oligomers is the tendency to form defects and traps at crystal grain boundaries over time.

2   Photoluminescence from Organic Materials

Electroluminescence and photoluminescence are related physical phenomena. Electroluminescence from organic semiconductor materials is the process whereby light is emitted from a layer of an organic material on passage of a current of electricity through it. In order to understand electroluminescence a brief description of photoluminescence first may be of help. Photoluminescence is the process whereby light is absorbed by the organic compound and then re-emitted, usually at a longer wavelength, see Figure 4.1. These materials are often referred to as being fluorescent or phosphorescent. The former is a faster process than the latter. Absorption of light by a conjugated organic molecule can give rise to an n–π^* or π–π^* transition and promote an electron to the LUMO. The energy levels are broadened by vibration and rapid vibronic relaxation to the bottom of the LUMO is followed by radiative decay to the ground state, see Figure 4.1. Hence, the photoluminescence absorption and emission spectra are broad bands, as shown in Figure 4.2, with the latter red-shifted to longer wavelengths. The electroluminescence and photoluminescence emission spectra of a particular organic material are almost always identical, since they originate from the same energy levels. The only difference is the method of excitation of an electron from the HOMO to the LUMO. In a molecular solid, intermolecular interactions and disorder also broaden the transitions. In conjugated organic molecules and polymers, emission is often in the visible region of the electromagnetic spectrum. The excited carriers are mobile: intermolecular excitation transfer occurs *via* neutral excitons or polarons in conjugated polymers. Excitons consist of bound electron-hole

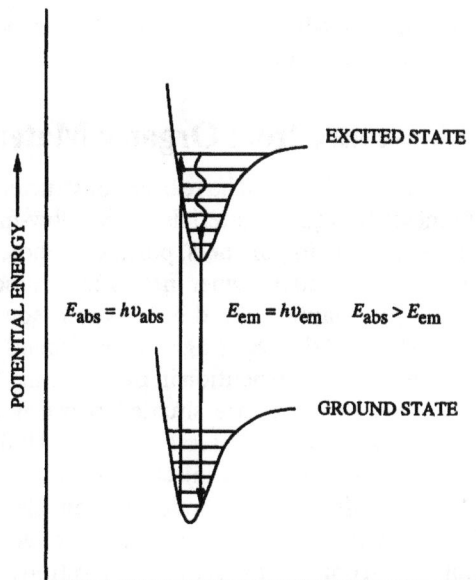

Figure 4.1 *Schematic representation of the mechanism of photoluminescence from a conjugated organic material. Absorption of a photon of a given energy can lead to promotion of an electron into one of the energy levels of the excited state. Relaxation from the lowest excited energy state with the highest population gives rise to emission at a lower frequency than that, v_{abs}, of the absorbed photon. The frequency, v_{em}, of light emitted of maximum intensity is related to the band gap, E_{em}, between the lowest energy levels of the ground state and excited state ($E = hv$) where h is Plank's constant). This is representative from the Franck–Condon principle.*

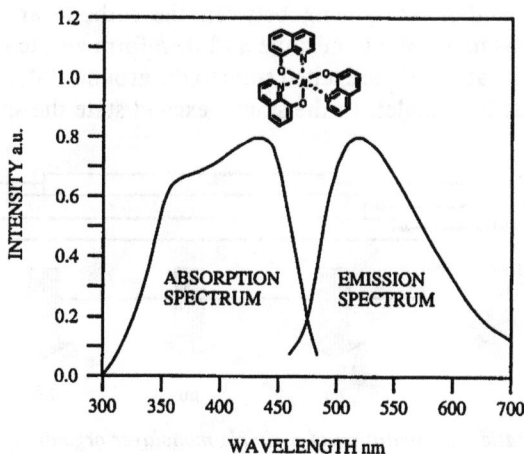

Figure 4.2 *Absorption and emission spectra of aluminium tris(quinolate) [Alq_3].*

pairs in the LUMO and HOMO, respectively, whereas polarons are single carriers screened by lattice relaxation.

3   Electroluminescence from Organic Materials

Organic light-emitting diodes (OLEDs) have been constructed using a variety of classes of electroluminescent organic materials, such as low-molar-mass organic compounds, oligomers, side-chain polymers, polymer blends, conjugated main-chain polymers and cross-linked polymer networks. These OLEDs have a similar configurations, see Chapters 5 and 6. However, some of the common features and characteristics of OLEDs using small molecules and polymers will be discussed here in order to avoid repetition in the following two chapters. One or more layers of organic material are situated between two electrodes of different materials and work functions, see Figure 4.3. Under a forward bias (voltage) radical anions are generated by injection of electrons into the organic material from a low-work-function cathode, *e.g.* an electrode made from aluminium or calcium. At the same time holes are injected as radical cations into the organic material from the anode, *i.e.* an electrode with a high work function, such as indium/tin oxide. The electrons and holes travel through the organic material in opposite directions towards the electrode of opposite polarity under the influence of the electric field. This construction creates a diode because the current flowing between the cathode and the anode under positive bias (voltage) is orders of magnitude higher than that flowing in the opposite direction under negative bias, *i.e.* if the direction of the electric field is reversed. The rectification ratio of monolayer and multilayer OLEDs using either electroluminescent small (low-molar-mass) molecules or LEPs is usually very high. Although very little current flows under negative bias and no light is emitted, AC driving, rather than DC driving, can lead to longer device lifetimes due to the reduction in the build-up of space charge at the interface between one of the electrodes and the adjacent organic material.

The electrons and holes flowing between the cathode and the anode in opposite directions may collide on route and then form an excited state, *i.e.* an exciton. Upon relaxation of the excited state to the ground state, light is emitted, if the excited state is a singlet. In the singlet excited state the spin states of the

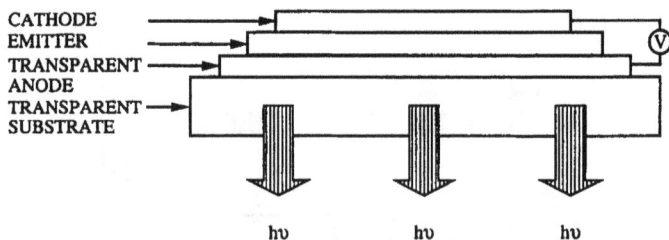

Figure 4.3   *Schematic representation of a simple monolayer organic light-emitting diode (OLED) incorporating an electroluminescent material between a transparent anode and a cathode.*

two electrons in the LUMO and HOMO energy levels are of opposite sign, whereas in the triplet excited state the spin states of the two electrons are of equal sign. Triplets usually relax to the ground state by a non-radiative pathway. However, since the probability of triplet formation to singlet formation for excitons produced by the recombination of a hole and an electron is 3:1, the maximum efficiency of electroluminescence from the excited states of organic materials is equal to 25%.

Some authors claim[42] that in polymers, recombination occurs between free charged polarons, so that the maximum efficiency corresponds to the internal quantum efficiency, η_{int} (*i.e.* the number of photons emitted for every electron injected), as given by

$$\eta_{int} = \gamma \eta_R \Phi_F \qquad (1)$$

where ϕ_F is the quantum efficiency of fluorescence, η_R is the efficiency of singlet formation and γ is a double charge injection factor. The latter parameter will also depend on the electrodes and the match of the electrode work functions with the energy levels of the adjacent organic material. However, a more relevant parameter is the external quantum efficiency, η_{ext}, as given by the following relationship:[41-43]

$$\eta_{ext} = \frac{\eta_{int}}{2n^2} \qquad (2)$$

where n is the refractive index of the organic material. Therefore, the external quantum efficiency is substantially lower than the internal quantum efficiency, since the refractive index of organic compounds is often relatively high, *e.g.* $n = 1.4$ for a conjugated organic polymer. This suggests that a lot of the light generated is lost within the device itself due to internal reflection and re-absorption leading primarily to non-radiative decay. Indeed most of the light is lost ($\geqslant 85\%$). The power efficiency, η_E, is the ratio of light output power to electrical input power and is related[44,45] to the operating voltage, V_{op}, and the energy of the emitted phonons, E_p in eV, by the equation

$$\eta_E = \eta_{int} \frac{E_p}{V_{op}} \qquad (3)$$

If the power efficiency, η_E, in lumens per watt (lm W^{-1}) is high, then the OLED is bright at low current density and applied low voltage. A high power efficiency generally corresponds to a more extended device lifetime than that of comparable devices with the same brightness, but higher operating voltages and power consumption. Low voltage and power consumption are also pre-requisites for battery-operated, portable devices. The luminous efficiency in lumens per watt (lm W^{-1}) is the product of the power efficiency, η_E, and the luminous efficacy, which compensates for the wavelength dependency of human eye sensitivity.[46] This is at a maximum for yellow-green light in the centre of the visible part of the electromagnetic spectrum (ca. 550 nm). The brightness of a display is also

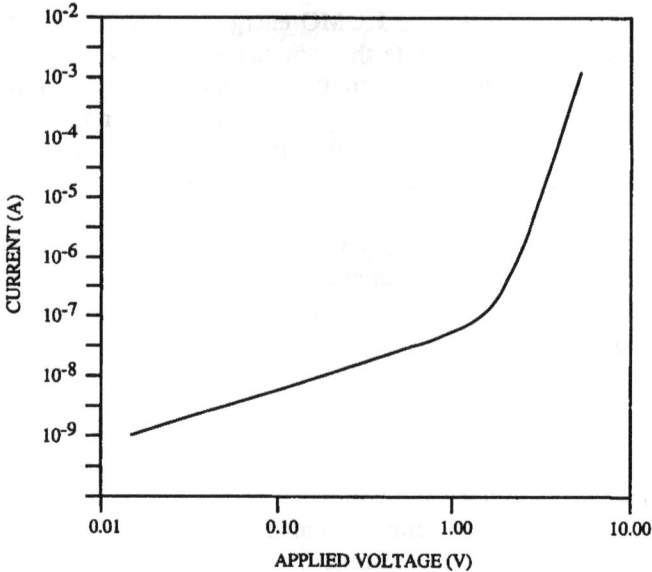

Figure 4.4 *Typical plot of the current,* I, *against voltage,* V, *of an organic light-emitting diode (OLED).*

often expressed in terms of the observed luminance in candelas per unit area (cd m^{-2}).

Figure 4.4 depicts a typical plot of the current, I, against voltage, V, for an OLED. A threshold voltage can be seen by a sharp change in the steepness of the I–V curve. The flow of electric current and, therefore, electroluminescence, takes place above this threshold voltage. The barriers to charge injection into the organic material must first be overcome by an appropriate electric field strength. The current density and intensity of light emitted then increase rapidly with increasing applied voltage. Plots of electric current *versus* voltage (I–V) curves are characteristic of a given OLED. It was found that multilayer OLEDs could significantly reduce operating voltage and power consumption, while, at the same time, increasing the luminous efficiency. In a bilayer OLED consisting of two layers of different organic materials, the LUMO of an electron transport layer (ETL) is matched to the work function of the cathode, while the HOMO of a hole transport layer (HTL) is matched to the work function of the anode. This leads to a more balanced injection of holes and electrons. Recombination then takes place at the interface between the two organic layers away from the electrodes where quenching can take place, see Figures 4.5 and 4.6 for the energy diagrams of monolayer and bilayer OLEDs, respectively. Recombination and light emission then takes place in either one of the layers near the organic layer interface.[47-51] If a third layer of emissive material is incorporated between the ETL and HTL layers to produce a three-layer OLED, then the power efficiency may be even higher and operating voltages lower. Electroluminescence then takes place in the central emission layer where the holes and electrons recombine.

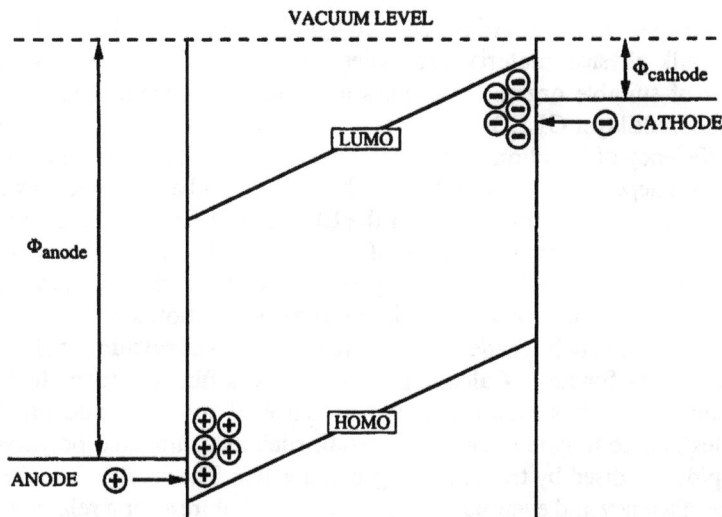

Figure 4.5 *Schematic energy level diagram of a generalised monolayer organic light-emitting diode (OLED).[34]*

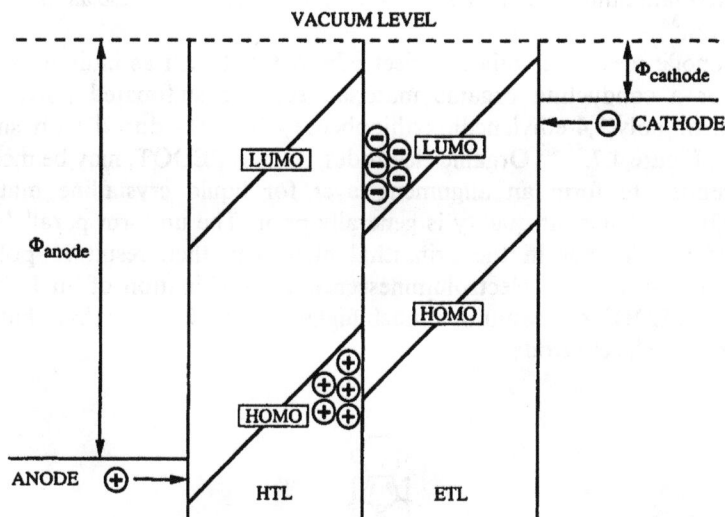

Figure 4.6 *Schematic energy level diagram of a generalised bilayer organic light-emitting diode (OLED).[34]*

It is evident from the above discussion that the threshold voltage, current density, power efficiency, luminous efficiency and, to some extent, device lifetime of OLEDs using organic low-molar-mass compounds, oligomers and polymers depends on intrinsic molecular properties, such as HOMO and LUMO energy levels, efficiency of hole and electron injection and subsequent transport, efficiency of singlet formation and fluorescence efficiency. The wavelength and bandwidth of the emitted light is also of critical importance. All of these bulk physical material properties can be influenced by the design and synthesis of suitable organic molecules and their subsequent combination in single and multilayer OLEDs.

The efficiency of injection of holes and electrons into the organic material in an OLED is dependent on the nature of the cathode and anode and how closely they match their work functions with the LUMO and HOMO, respectively, of the organic material in the centre of the device. There is some degree of flexibility of device design with respect to matching these parameters by judicious choice of electrode materials and organic compounds.

The cathode is usually made of a reactive metal such as calcium or aluminium with a low work function. Calcium has a lower work function than aluminium, but aluminium is less reactive in the presence of moisture and air. Metal electrodes and conjugated organic material, such as conjugated polymers, are both rapidly oxidised by traces of oxygen and water. The result is a rapid decay in device efficiency and eventually complete device failure after a relatively short time.[1,52–54] Therefore, effective encapsulation of OLEDs is generally required in order to exclude oxygen and water and produce acceptable device operational lifetimes. Other metals, such as gold and mixed metal/metal oxides, such as aluminium/aluminium oxide, have also been investigated for use as the cathode in OLEDs.[38]

The anode may be a standard electrode material such as indium/tin oxide (ITO) or a conducting organic material, such as sulfonated poly(aniline) (PANi), or poly(3,4-ethylenedioxythiophene) (PEDOT) doped with sulfonic acid, see Figure 4.7.[55–60] Organic electrodes, such as PEDOT, may be mechanically rubbed to form an alignment layer for liquid crystalline materials, although the alignment quality is generally poor. The uniform parallel alignment of the director in the azimuthal plane can then result in polarised photoluminescence and electroluminescence. A combination of an ITO electrode and PEDOT can result in a much higher power efficiency than that using ITO alone as the electrode.

poly(3,4-ethylenedioxythiophene): PEDOT

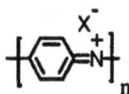

poly(aniline) salt: PANi with counterion

Figure 4.7 *Molecular structure of two typical organic electrode materials.*

OLEDs can be addressed in a similar fashion to LCDs, see Chapter 2, *i.e.* directly with segmented electrodes, see Figure 4.8, by multiplexed addressing with rows and columns of electrode strips, see Figure 4.9 and by active matrix addressing with one transistor at each pixel, see Figure 4.10. The major

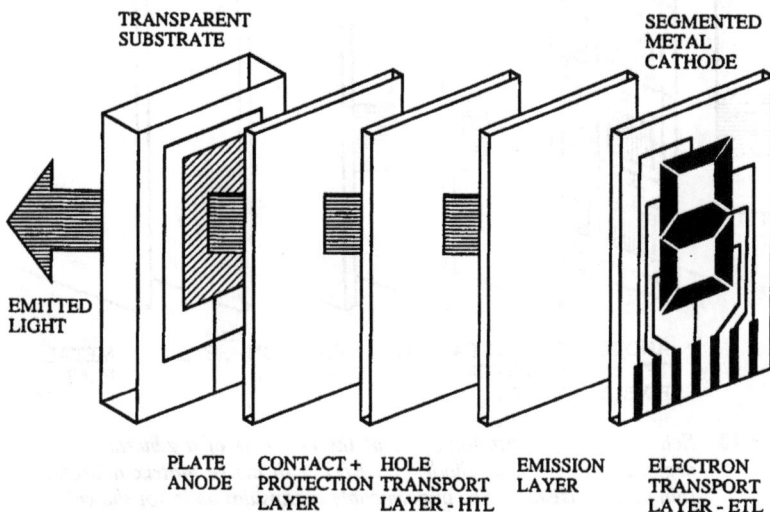

| PLATE
ANODE | CONTACT +
PROTECTION
LAYER | HOLE
TRANSPORT
LAYER - HTL | EMISSION
LAYER | ELECTRON
TRANSPORT
LAYER - ETL |

Figure 4.8 *Schematic representation some of the elements of a generalised multilayer organic light-emitting diode (OLED) with direct addressing. The thin metallic cathode segments are connected directly to the electron-transport layer (ETL). The impermeable encapsulation is not shown.*

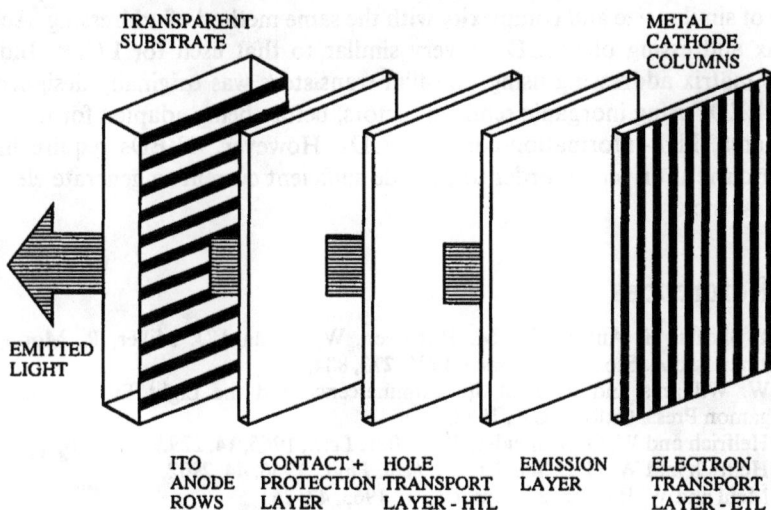

| ITO
ANODE
ROWS | CONTACT +
PROTECTION
LAYER | HOLE
TRANSPORT
LAYER - HTL | EMISSION
LAYER | ELECTRON
TRANSPORT
LAYER - ETL |

Figure 4.9 *Schematic representation some of the elements of a generalised multilayer organic light-emitting diode (OLED) with multiplex addressing. The thin metallic cathode strips are connected directly to the electron-transport layer (ETL). The impermeable encapsulation is not shown.*

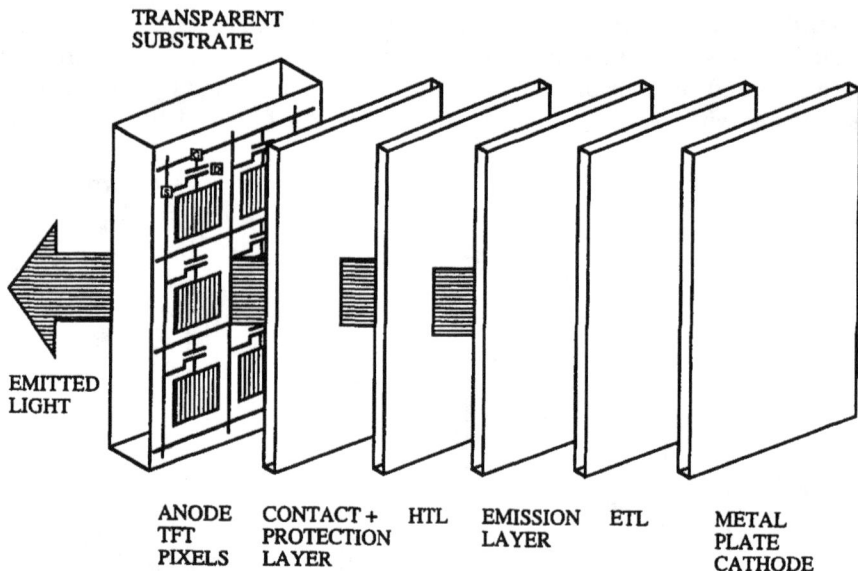

TRANSPARENT
SUBSTRATE

EMITTED
LIGHT

ANODE	CONTACT +	HTL	EMISSION	ETL	METAL
TFT	PROTECTION		LAYER		PLATE
PIXELS	LAYER				CATHODE

Figure 4.10 *Schematic representation some of the elements of a generalised multilayer organic light-emitting diode (OLED) with active matrix addressing using thin film transistors. The impermeable encapsulation is not shown.*

difference between the addressing of LCDs and OLEDs is that a current of several mA cm^{-2} flows between the electrodes in an OLED. Therefore, the power consumption is significantly higher for OLEDs than that of a comparable LCD of similar size and complexity with the same method of addressing. Active matrix addressing of OLEDs is very similar to that used for LCDs. Indeed active matrix addressing using thin-film transistors was originally designed to drive LEDs using inorganic semiconductors, before being adapted for use with large-area, high-information-content LCDs. However, OLEDs require high-power current drivers in order to provide sufficient current to generate electro-luminescence.

4 References

1 J. R. Sheats, H. Antoniadis, M. Hueschen, W. Leonard, J. Miller, R. Moon, D. Roitman and A. Stocking, *Science*, 1996, **273**, 884.
2 E. W. Williams and R. Hall, in 'Luminescence and the Light Emitting Diode', Pergamon Press, Oxford, UK, 1978.
3 W. Helfrich and W. G. Schneider, *Phys. Rev. Lett.*, 1965, **14**, 229.
4 W. Helfrich and W. G. Schneider, *J. Chem. Phys.*, 1966, **44**, 2902.
5 W. Mehl and W. Bucher, *Z. Phys. Chem.*, 1965, **47**, 76.
6 D. F. Williams and M. Schadt, *Proc. IEEE*, 1970, **58**, 476.
7 J. Dresner, *RCA Rev.*, 1969, **30**, 322.
8 R. E. Kellogg, *J. Chem., Phys.*, 1966, **44**, 411.
9 P. S. Vincett, W. A. Barlow, R. A. Hann and G. G. Roberts, *Thin Solid Films*, 1982, **94**, 171.

10 F. J. Campas and M. Goutermann, *Chem. Phys. Lett.*, 1977, **48**, 233.

11 J. Kalinowski, J. Godlewski and Z. Dreger, *Appl. Phys. Lett. A.*, 1985, **37**, 179.

12 C. W. Tang and S. A. Van Slyke, *Appl. Phys. Lett.*, 1987, **51**, 913.

13 C. W. Tang, S. A. Van Slyke and C. H. Chen, *Appl. Phys. Lett.*, 1989, **65**, 3610.

14 M. Era, C. Adachi, T. Tsutsui and S. Saito, *Chem. Phys. Lett.*, 1991, **178**, 488.

15 C. Adachi, S. Tokito, T. Tsutsui and S. Saito, *Jpn. J. Appl. Phys.*, 1988, **27**, L713.

16 C. Adachi, T. Tsutsui and S. Saito, *Appl. Phys. Lett.*, 1989, **55**, 1489.

17 C. Adachi, T. Tsutsui and S. Saito, *Appl. Phys. Lett.*, 1990, **56**, 799 and 1990, **57**, 531.

18 C. Adachi, T. Tsutsui and S. Saito, *Acta Polytech. Scand. Appl. Phys.*, 1990, **170**, 145 and 215.

19 Y. Hamada, C. Adachi, T. Tsutsui and S. Saito, *Jpn. J. Appl. Phys.*, 1992, **31**, 1812.

20 A. Dodabalapur, L. J. Rothberg and T. M. Miller, *Appl. Phys. Lett.*, 1994, **65**, 2308.

21 J. Kido, M. Kimura and K. Nagaim, *Science*, 1995, **67**, 1332.

22 S. Miyata and H. S. Nalwa, in 'Organic Electroluminescent Materials and Devices', Gordon and Breach, New York, USA, 1997.

23 K. Y. Law, *Chem. Rev.*, 1993, **93**, 449.

24 H. Naarmann, in 'Conjugated Polymer Materials: Opportunities in Electronics, Optoelectronics and Molecular Electronics', Eds. J. L. Bredas, and R. R. Chance, Kluwer, Dordrecht, Netherlands, 1990.

25 A. Bohnen, H. J. Räder and K. Müllen, *Synth. Met.*, 1992, **47**, 37.

26 N. F. Colaneri, D. D. C. Bradley, R. H. Friend, P. L. Burn, A. B. Holmes and C. W. Spangler, *Phys. Rev.*, 1990, **B42**, 11670.

27 T. J. Marks and M. A. Ratner, *Angew. Chem.*, 1995, **107**, 167.

28 M. Baumgarten and K. Müllen, in 'Topics in Current Chemistry', Springer Verlag, Berlin, Germany, 1994, Vol. 169.

29 R. H. Friend, D. D. C. Bradley and A. B. Holmes, *Phys. World*, 1992, **5/(11)**, 42.

30 D. D. C. Bradley, *Synth. Met.*, 1993, **54**, 401.

31 R. H. Friend and N. C. Greenham, *Solid State Phys.*, 1995, **49**, 1.

32 D. D. C. Bradley, *Curr. Opin. Solid State Mater. Sci.*, 1996, **1**, 789.

33 R. W. Gymer, *Endeavor*, 1996, **20**, 115.

34 J. Salbeck, *Ber. Bunsenges. Phys. Chem.*, 1996, **100**, 1667.

35 L. J. Rothberg and A. J. Lovinger, *J. Mater. Res.*, 1996, **11**, 3174.

36 K. Kaneto, K. Yoshino, K. Koa and Y. Inuishi, *Jpn. J. Appl. Phys.*, 1974, **18**, 1023.

37 R. H. Partridge, *Polymer*, 1983, **24**, 755.

38 J. H. Burroughs, D. D. C. Bradley, A. R. Brown, N. Marks, K. Mackay, R. H. Friend, P. L. Burn and A. B. Holmes, *Nature*, 1990, **347**, 539.

39 J. H. Burroughes, D. D. C. Bradley, A. R. Brown, R. N. Marks, R. H. Friend, P. L. Burn and A. B. Holmes, *Nature*, 1990, **347**, 539.

40 F. Wudl, P. M. Allemand, G. Srdanov, Z. Ni and D. McBranch, *ACS Symp. Ser.*, 1991, 455.

41 D. Braun and A. J. Heeger, *Appl. Phys. Lett.*, 1991, **58**, 1982.

42 A. J. Heeger, in 'Primary Photoexcitation in Conjugated Polymers', Ed. N. S. Sacricifiri, World Scientific Press, New York, USA, 1987, Chapter 2.

43 N. C. Greenham, R. H. Friend and D. C. C. Bradley, *Adv. Mater.*, 1994, **6**, 491.

44 T. Tsutsui and S. Saito, *NATO ASI Series*, 1993, 246.

45 S. Tasch, W. Graupner, G. Leising, L. Pu, W. Wagner and R. H. Grubbs, *Adv. Mater.*, 1995, **7**, 903.

46 S. A. Carter, M. Angelapoulos, S. Karg, P. J. Brock and J. C. Scott, *Appl. Phys. Lett.*, 1997, **70**, 2067.

47 J. C. Carter, I. Grizzi, S. K. Heeks, D. J. Lacey, S. G. Latham, P. G. May, O. R. de los Panos, K. Pichler, C. R. Towns and H. F. Wittmann, *Appl. Phys. Lett.*, 1997, **71**, 34.

48 N. C. Greenham, R. H. Friend, A. R. Brown, D. C. C. Bradley, K. Pichler, P. L. Burn, A. Kraft and A. B. Holmes, *Proc. SPIE Int. Soc. Opt. Eng.*, 1993, **1910**, 84.

49 A. R. Brown, D. C. C. Bradley, J. H. Burroughes, R. H. Friend, N. C. Greenham, P. L. Burn and A. B. Holmes, *Appl. Phys. Lett.*, 1992, **61**, 2793.

50 Y. Yang and Q. Pei, *J. Appl. Phys.*, 1995, **77**, 4807.
51 E. Buchwald, M. Meier, S. Karg, P. Pösch, H.-W. Schmidt, P. Strohriegel, W. Riess and M. Schwoerer, *Adv. Mater.*, 1995, **7**, 839.
52 R. D. Scurlock, B. Wang, P. R. Ogilby, J. R. Sheats and R. L. Clough, *J. Am. Chem. Soc.*, 1995, **117**, 10194.
53 T. Zyung and J.-J. Kim, *Appl. Phys. Lett.*, 1995, **67**, 3420.
54 K. Z. Xing, N. Johansson, G. Beamson, D. T. Clark, J.-L. Brédas and W. R. Salaneck, *Adv. Mater.*, 1997, **9**, 1027.
55 A. C. Fou, O. Onitsuka, M. Ferreira, D. Howie and M. F. Rubner, *Polym. Mater. Sci. Eng.*, 1995, **72**, 160.
56 A. C. Fou, O. Onitsuka, M. Ferreira, M. F. Rubner and B. R. Hsieh, *Mater. Res. Soc. Symp. Proc.*, 1995, **369**, 575.
57 M. Onoda and K. Yoshino, *Jpn. J. Appl. Phys.*, 1995, **34**, L260.
58 M. Onoda and K. Yoshino, *Jpn. J. Appl. Phys.*, 1995, **78**, 4456.
59 A. C. Fou, O. Onitsuka, M. Ferreira, M. F. Rubner and B. R. Hsieh, *J. Appl. Phys.*, 1996, **79**, 7501.
60 M. Ferreira, O. Onitsuka, A. C. Fou, B. R. Hsieh and M. F. Rubner, *Mater. Res. Soc. Symp. Proc.*, 1996, **413**, 49.

Organic Light-Emitting Diodes Using Low-Molar-Mass Materials (LMMMs)

1 Introduction

The first prototype organic light-emitting diodes (OLEDs) using low-molar-mass organic and organometallic compounds, such as metal chelates, were constructed over 30 years ago.[1,2] However, despite intense research and development and a range of attractive features, OLEDs have only started to penetrate the market for flat panel displays at the beginning of the 21st century.[3]

Electroluminescence from organic materials, as opposed to doped inorganic semiconductors, was first discovered using single crystals of anthracene (1) (Table 5.1) immersed in an electrochemical cell.[1,2] The cell contained an electrolytic solution of negatively charged anthracene ions, prepared from a solution of anthracene and sodium in tetrahyrofuran next to the cathode and a solution of positively charged anthracene ions, prepared from a solution of anthracene and aluminium trichloride in nitromethane, next to the anode. The two liquid electrolytic solutions were physically separated from each other by the solid anthracene crystal. The two electrodes were connected to a source of potential. The electrolytic solutions were responsible for charge transport to the electroluminescent anthracene crystal of a current of electrons injected from the cathode and holes injected from the anode, respectively.[1-5] A visible emission of blue light ($hv < 3.1$ eV) was observed above a certain threshold voltage from the side of the device under the action of an electric current through the cell. This was the first unambiguous report of electroluminescence from an organic material caused by the recombination of holes and electrons to form an emissive singlet exciton. Since anthracene is an insulating organic material with a high resistivity and a low permitivity ($\varepsilon = 3.4$), the maximum electron mobility through anthracene is still low ($\mu_e \approx 0.4\,\mathrm{cm^2\,V^{-1}\,sec^{-1}}$) compared with that through metals or doped graphite, although relatively high for an organic material of this type. Consequently, the electric current through anthracene is

Table 5.1 *Glass transition temperature (°C), emission maximum (nm) and colour for the compounds (1–3)*

Molecular structure	Acronym	T_g	λ_{max}	Colour	Ref
1			430	Blue	1, 2
2	TPAC	60			13
3	Alq$_3$	175	530	green	14

very low or zero at low field strengths. High voltages ($V_{op} \approx$ 400–2000 V) had to be applied in order to create the high electric fields ($E \approx 10^4 \sim 10^5 \, \text{V cm}^{-1}$) necessary to induce an electric current density of a significant magnitude ($I \approx 10^{-5} \, \text{A cm}^{-2}$) to generate visible blue electroluminescence from the crystals. This was due to the thickness of the crystals used ($d \approx 5 \, \text{mm}$), since in the high current regime

$$E \propto \frac{V}{L} \tag{1}$$

The electroluminescence was located in the part of the anthracene crystal next to the hole-injecting anode, which suggests an imbalance of charge-carrier injection and transport. The intensity of light was linearly proportional to the magnitude of the current above the threshold voltage for electroluminescence. This appeared to correspond to a space-charge limited (SCL) model of charge-carrier injection and transport. The external quantum efficiency of these simple, prototype OLEDs was reported as being remarkably high ($\eta_{ext} \approx 35\%$). The mobility of both holes and electrons in crystalline anthracene is high for an organic material. Therefore, recombination may also occur in the bulk of the material. This may partially explain the high quantum efficiency of this device, which corresponds to a photoluminescence efficiency of 99%.[6]

Considerably lower operating voltages were found using solid, metallic electrodes instead of the two liquid electrolytic solutions used for charge transport and injection to thin planar wafers of cleaved anthracene crystals.[7] This prototype, solid-state OLED was glued to a semi-transparent solid substrate for mechanical stability and planarity. The cathode was a negatively-charged alkali metal salt [sodium (Na), potassium (K), or a sodium–potassium (Na-K) alloy] of anthracene with a reasonably low work function ($\phi = 2.05\,eV$ for the Na–K alloy and $\phi = 2.3\,eV$ for K). The cathode was prepared by reaction of the metal with the interface with the anthracene wafer. The anode was made from gold (Au) or an alloy of equal amounts of selenium and tellurium ($Se_{50}-Te_{50}$), both of which possess a high work function ($\phi = 5.2\,eV$ for gold (Au) and $\phi = 4.8\,eV$ for $Se_{50}-Te_{50}$). Relatively low threshold voltages ($V_{th} > 35\ V$) and operating voltages ($V_{op} > 100\ V$) for visible electroluminescence of blue light ($\lambda_{max} = 430\,nm \approx 3\,eV$) were obtained using the very thin wafers of anthracene ($30\,\mu m$ to $200\,\mu m$) compared to those (400–$2000\ V$) observed using thick ($5000\,\mu m$) anthracene crystals.[1,2] This is a clear function of the difference in the thickness, d, of the organic electroluminescent material.

The electroluminescence intensity was linearly proportional to the current and almost independent of the sample thickness above the threshold voltage. The external quantum efficiency ($1\% < \eta_{ext} < 8\%$) of these OLEDs using thin wafers of anthracene appeared to be lower than that observed for the OLEDs using anthracene crystals ($\eta_{ext} \approx 35\%$). This may be attributable in part to the mismatch of the work functions of the electrodes and the LUMO and HOMO energy levels of anthracene. Remarkably high current densities ($I = 10^{-1}\,A$ cm^{-2}) were observed for high voltages ($V_{op} = 600$ V) without the onset of dielectric breakdown of the diode or apparent electrochemical degradation of the organic anthracene layer.

Very thin layers ($3\,\mu m < d < 30\,\mu m$) of anthracene were produced by deposition of anthracene powder followed by melting and subsequent recrystallisation of the layer.[7] However, the presence of pinholes in extremely thin layers meant that thicker layers ($d > 14\,\mu m$) had to be used. However, very thin layers ($d \approx 0.1\,\mu m$) without pinholes were finally produced by physical vapour-deposition of anthracene under high vacuum conditions.[8-10] OLEDs produced using these layers exhibited low threshold voltages ($V_{th} \approx 3.5$ V) and operating voltages ($V_{op} < 30\ V$) using solid electrodes. However, the quantum efficiency of these devices was low ($\eta_{ext} \approx 0.05\%$), probably due to the poor quality of the films deposited.

The configuration and construction of monolayer and multilayer OLEDs have undergone substantial changes and modifications since these first reports of organic electroluminescence from low-molar-mass materials. Several types of OLEDs using small organic and organometallic molecules are described schematically below.

2 Monolayer Organic Light-Emitting Diodes Using LMMMs[1-9]

Display Configuration

A thin film (≈ 100 nm) of the organic electroluminescent material is sandwiched between two electrodes supported on a solid substrate, such as glass, see Figure 5.1. The device substrate providing the mechanical support to the anode, electroluminescent polymer layer and the cathode is transparent or semi-transparent in order to allow the light generated by electroluminescence to actually leave the display and be observed. An appropriate voltage can be applied between the electrodes. The cathode, *e.g.* a thin layer ($\approx 7-15$ nm) of aluminium, calcium or a magnesium/silver alloy, has a low work function (ionisation potential), whereas the anode, *e.g.* a thin metallic layer (15 nm) of indium–tin oxide (ITO), has a high work function. No light is emitted in the absence of an electric current and so the non-activated pixels appear dark. The application of a voltage between the two electrodes above the threshold voltage, V_{th}, leads to the injection of electrons from the cathode and holes from the anode into the organic electroluminescent material and the emission of light. The threshold voltages and operating voltages for monolayer OLEDs using small molecules are primarily determined by the thickness of the organic layer, the compatibility of the work functions of the electrodes and the LUMO and HOMO of the organic material.

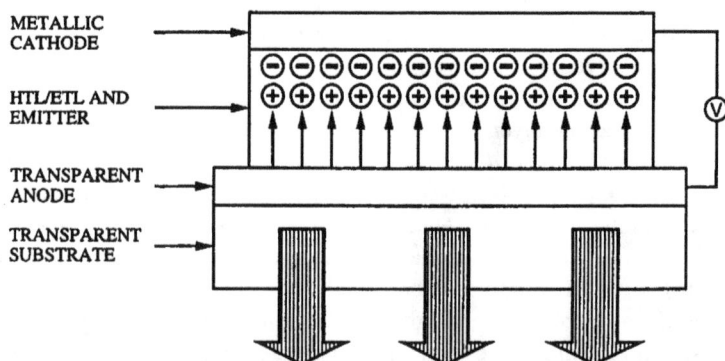

Figure 5.1 *Schematic representation of a monolayer OLED using low-molar-mass materials incorporating an electroluminescent material between a transparent anode and a cathode.*

3 Bilayer OLEDs Using LMMMs

The external quantum efficiency of electroluminescent devices using one layer of organic light-emitting material sandwiched between two metallic electrodes was generally found to be very low, whatever the organic material used. This phenomenon has several causes intrinsic to the device configuration. If there is a difference between the injection barriers of charged species at the electrodes then one type of charge-carrier will be preferentially injected. Furthermore, a given organic material will transport one type of charge-carrier preferentially, i.e. the mobility of holes and electrons through the organic monolayer will be different, often substantially so. This will lead to a concentration of both charge-carriers near the electrode injecting the minority charge-carrier. Therefore, recombination of holes and electrons, consequent exciton formation and subsequent radiative decay of singlets takes place in a narrow planar region near the interface between one electrode and the organic material. However, the metal atoms of the electrode generally induce non-radiative quenching of the excitons. Non-radiative transfer of energy from the excitons to the metal, *e.g.* by Förster transfer, can also occur near to the interface. If the ratio between the two charge-carrier currents is large then a number of one of the charge carriers can pass from one electrode to the other without recombination with the oppositely charged species being able to take place. A combination of all these phenomena generally leads to a low external quantum efficiency for single-layer OLEDs.

Consequently, interest in organic electroluminescence waned for many years until the report by Tang and Van Slyke (Eastman Kodak Corporation) of bilayer OLEDs with a much higher efficiency and lower operating voltage due, at least in part, to a more balanced charge injection from the anode and cathode.[11,12] The emission is also concentrated at the interface between the two organic layers away from the metallic electrodes, which reduces quenching by the metal atoms. There are several possibilities for constructing bilayer OLEDs. These include an electron-transport layer (ETL) and a combined hole-transport (HTL) and emission layer. Conversely a hole-transport layer and a combined electron-transport and emission layer is also effective.

Display Configuration

A thin film ($\approx 100\,\text{nm}$) of an organic hole-transport layer (HTL) supports a second thin film of an organic electron-transport layer (ETL) situated between two electrodes supported on a substrate, see Figures 5.2 and 5.3. One or both of the layers form the emission layer. The anode is transparent in order to allow the passage of the light generated. The electrodes are connected to a source of potential. The metal cathode has a low work function, which is matched as closely as possible to the LUMO of the ETL in order to facilitate efficient electron injection. The HOMO of the HTL is also matched as near as possible to the high work function of the anode, *e.g.* indium–tin oxide (ITO) or an organic polymer electrode such as poly(aniline) salt (PANi). No light is emitted in the

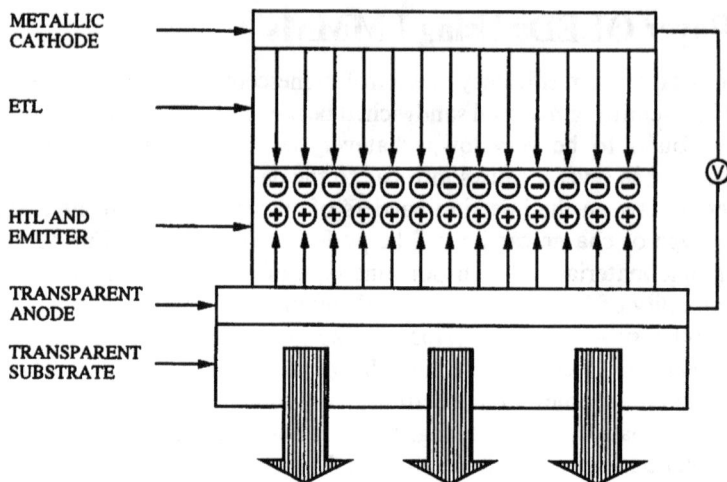

Figure 5.2   *Schematic representation of a bilayer OLED using low-molar-mass materials incorporating a combined hole-transport and emission layer and an electron-transport layer situated between a transparent anode and a cathode.*

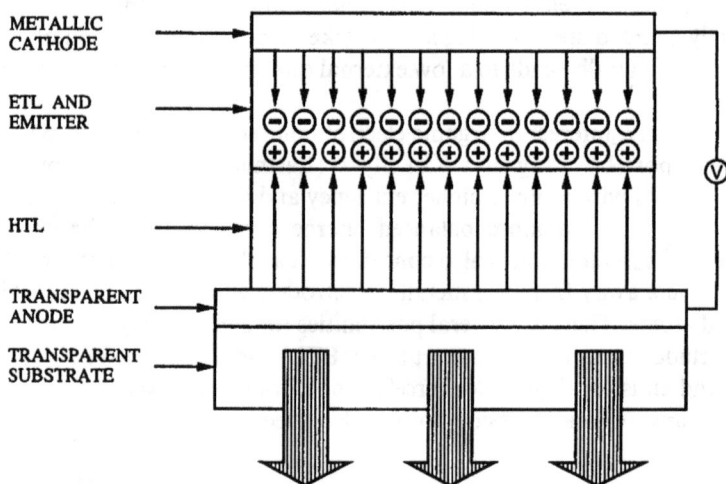

Figure 5.3   *Schematic representation of a bilayer OLED using low-molar-mass materials incorporating a hole-transport layer and a combined electron-transport and emission layer situated between a transparent anode and a cathode.*

absence of an electric current and so the non-activated pixels appear dark. Application of a voltage above the threshold voltage leads to injection of electrons from the cathode into the ETL and holes from the anode into the HTL. The close matching of the work functions of the individual electrode materials to those of the ETL and the HTL lowers the barriers to charge injection. Thus, the threshold voltage is lower than that of analogous monolayer OLEDs. The resultant electric current passing through the organic layers is much more balanced than that found in monolayer OLEDs, since each layer preferentially transmits one type of charge-carrier and blocks the transport of the oppositely charged species. Thus, a high percentage of the injected electrons and holes congregate at the interface between the ETL and HTL, where charge recombination and exciton formation can take place. This may further lower the barrier to charge injection at the electrodes. The concentration of the charge-carriers at the interface between the HTL and ETL gives rise to the efficient emission of light from the excitons in a thin strip in one or both of the layers, either side of the interface, depending on the presence or absence of emitters in the individual layers, see Figures 5.2 and 5.3. Thus, the activated pixels emit light and appear bright.

This bilayer OLED configuration generally results in a relatively high quantum efficiency ($1\% < \eta_{ext} < 4\%$) and high luminance at low operating voltages. Tang and van Slyke reported[11] a green monochrome, bilayer OLED with high brightness at low operating voltages using a positive bias across a bilayer of the diamine (2)[13] as the HTL and Alq₃ (3)[14] as the ETL and emission layer, see Tables 5.1 and 5.2. The metal chelate Alq₃ (3) has a relatively high electron-transport mobility ($\mu_e = 10^{-5} \, cm^2 \, V^{-1} \, s^{-1}$) for an organic material, whereas the diamine (2) has a high hole-transport mobility ($\mu_h = 10^{-3} \, cm^2 \, V^{-1} \, s^{-1}$). The cathode was made from an alloy (Mg_{10}–Ag_1) of magnesium (Mg) and silver (Ag) with a relatively high work function ($\phi = 3.6 \, eV$) and the anode consisted of a layer of indium–tin oxide (ITO) with a high work function ($\phi = 4.6 \, eV$). The advantage of a Mg–Ag alloy as the cathode is that, despite the high work function, it is stable with respect to electrochemical oxidation in air. Very high values for the magnitude of the brightness ($L > 1000 \, cd \, m^{-2}$) can be achieved using higher operating voltages ($V_{op} > 10 \, V$). However, the half-life of

Table 5.2 *Configuration and typical electro-optical performance of an early bilayer OLED[11,12] using the low-molar-mass diamine (2) as HTL[13] and Alq₃ (3) as the ETL and emission layer[14]*

Property	Value
Cathode	Mg_{10}–Ag_1
Anode	ITO
Hole transport layer (HTL)	(2) 75 nm
Electron transport layer (ETL) and emission layer	(3) 60 nm
Threshold voltage, V_{th}	2.5 V
Operating voltage, V_{op}	5.5 V
Quantum efficiency, ϕ	1%
Power conversion efficiency	0.46%
Luminous efficiency	$1.5 \, lm \, W^{-1}$
Current density, I	10^{-1}–$10^2 \, mA \, cm^{-2}$
Colour	Green
Emission maximum, λ_{max}	550 nm
Bandwidth	200 nm
Brightness, L	$100 \, cd \, m^{-2}$

emission was found to be short (100 h) under normal drive conditions ($L = 50$ cd m^{-2} and $I = 5$ mA cm^{-2}). This was attributed to the degradation of the electrodes as manifested by the formation of black spots, although the exact mechanism of degradation was not completely understood at the time. All the organic and metallic layers are created by vacuum deposition at high vacuum (10^{-5} Torr).

4   Trilayer OLEDs Using LMMMs[15–22]

Even higher external quantum efficiencies were found for three-layer devices.[15–22] There are several reports of trilayer OLEDs using three distinct layers for electron transport, hole transport and emission.[15–20] This device configuration has the advantage that each layer can be optimised for one particular function, *i.e.* hole-transport, electron-transport or charge capture, recombination, exciton formation and subsequent emission of light. Simultaneous emission of red, green and blue light from a trilayer OLED can also be used to generate white light emission.[21,22] However, the fabrication of OLEDs with three distinct organic layers of controlled thickness, integrity and homogeneity with high production yield and low overall cost is a challenging task.

Display Configuration

A thin layer (≈ 100 nm) of a hole-transporting material with a low ionisation potential is situated next to the anode. An electron-transporting layer with a high electron affinity is situated next to the cathode. An organic emission layer is sandwiched between the HTL and the ETL, see Figure 5.4. At least one of the two electrodes is transparent in order to allow the passage of the light generated. A potential can be applied between the electrodes. The low work function of the metal cathode is matched as closely as possible to the LUMO of the ETL in order to facilitate efficient electron injection. The high work function of the anode is also matched as near as possible to the HOMO of the HTL. The electroluminescent material in the centre of the display is chosen to exhibit a high quantum efficiency of emission at the desired wavelength. No light is emitted in the absence of an electric current and so the non-activated pixels appear dark. Upon application of a voltage above the threshold voltage electrons are injected from the cathode into the ETL and holes are injected from the anode into the HTL. Therefore, a current of electrons passes through the ETL and a current of holes travels in the opposite direction through the HTL. This creates a concentration of oppositely charged species in the emissive layer. The electrons and holes then recombine in the emission layer or at the interface between the emission layer and the HTL and ETL. This leads to efficient charge recombination in the middle of the device with consequent generation of singlet excitons, from which light is then emitted, as well as formation of excitons in the generally non-emissive triplet state.

Very thin emitter layers (5 nm) are sufficient to produce bright electroluminescence.[19,21] However, even one or two layers of an organic electroluminescent

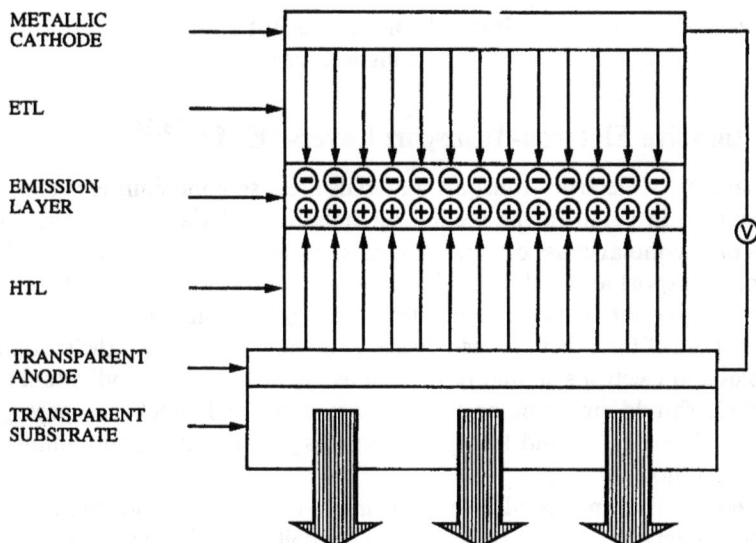

Figure 5.4   *Schematic representation of a trilayer OLED using low-molar-mass materials incorporating a hole-transport layer (HTL), an electron-transport layer (ETL) and a central emission layer situated between a transparent anode and a cathode.*

material deposited by Langmuir–Blodgett techniques situated between a HTL and ETL in a trilayer OLED were found to generate electroluminescence of a sufficient intensity to be visible to the eye.[15]

5   Low-Molar-Mass Organic Materials for OLEDs[23–25]

The low-molar-mass materials (LMMMs) designed and synthesised for use in OLEDs must satisfy a number of essential requirements in order to be of use in practical commercial devices. They must be capable of being deposited as a pure, uniform, thin solid film by vapour deposition under high vacuum. The individual layers should exhibit a high glass transition temperature in order to inhibit crystallisation during the lifetime of the device. They must be thermally, chemically, photochemically and electrochemically stable. They should function as a hole-transport layer, an electron-transport layer, a highly efficient fluorescent/phosphorescent emitter or a combined transport and emission layer depending on the device combination. Materials, which satisfy some or all of these requirements, are described in detail below. Where available the colour and maximum wavelength (λ_{max}) of electroluminescent emission of these materials are given in the Tables 5.1–5.12. Unfortunately, these values are not always readily available, especially from sources reporting the synthesis of new materials. The photoluminescence and electroluminescence spectra of organic or organometallic molecules are usually broad due to molecular vibrations, see

Chapter 4. Therefore, the observed colour does not always correspond exactly to the λ_{max}, as is the case for atomic emission spectra.

Non-Emissive Electron-Transport Layers (ETLs)[23,24]

The majority of electron-transport materials are also good emitters. These will be described in detail in the section on emissive materials. However, several classes of low-molar-mass organic materials are used as non-emissive, combined electron-transport and hole-blocking layers.[23,24] It is important that a charge-transport layer conducts predominantly one type of charge-carrier and blocks the transport of the other charged-species, *i.e.* the ratio between the mobility of the holes and electrons should be as high as possible for each individual layer. The HTL should predominantly transport holes and block the passage of electrons. The ETL should transport electrons preferentially and minimise the transport of holes.

Some typical examples of low-molar-mass, electron-transporting and hole-blocking materials for multilayer OLEDs are collated in Table 5.3. A *bis*(benz-imidazolyl)perylenedicarboximide derivative (4; R = H) was used in the first prototype OLED[25] using an emissive HTL and a non-emissive ETL. Electron-deficient nitrogen heterocycles are used as the ETL in OLEDs using LMMMs due to their high electron affinity attributable to the presence of the electro-negative nitrogen atoms in the five-membered heterocyclic ring. These include 2,5-disubstituted-1,3,4-oxadiazoles (5–7)[26,27] and triazoles (8). The 2-(4-*tert*-butylphenyl)-5-(biphenyl-4-yl)-1,3,4-oxadiazole (5), often referred to by its acronym of PBD, is used very often as a non-emissive ETL. However, compounds incorporating two oxadiazole rings, such as 7[28] are also used due to the lower tendency of thin layers containing them to crystallise in devices. 2,5-Disubstituted-1,3,4-triazoles such as compound 8[29] also contain a five-membered heterocyclic ring. However, triazoles contain three nitrogen atoms rather than the two nitrogen atoms of oxadiazoles, and consequently, triazoles generally possess a greater electron affinity than that exhibited by the analogous 2,5-disubstituted-1,3,4-oxadiazoles. The use of *bis*(benzimidazolyl)-naphthalenedicarboximides (9)[24] and thiopyran sulfones (10)[24] as ETLs has also been reported. The spiro-PBD (11) forms a stable glass with a high glass transition temperature due to its twisted structure as a result of the sp^3 bonding at the point of linking.[30–32] Although the melting points of these spiro-linked compounds are generally high due to their high molecular weight, they form the glassy state on quenching to temperatures below the melting point.[31,32]

Non-Emissive Hole-Transport Layers (HTLs)[23,24]

Since a large number of low-molar-mass emitters are also efficient electron-transport materials, there is a clear requirement for efficient, non-emissive, hole-transporting, electron-blocking LMMMs. Fortunately, a large range of hole-transporting, low-molar-mass organic materials had already been developed as

Table 5.3 *Molecular structure of the low-molar-mass, non-emissive ETL materials (4–11)*

Molecular structure	Acronym	Ref
4	PV	25
5	PBD	26–28
6	BND	26–28
7	OXD 1	28
8	TAZ	29
9	PD	24
10	TPS	24
11	Spiro-PDP	30, 31

photovoltaic materials for use in xerography and laser printing.[24] A whole series of low-molecular-weight, aromatic diamines have been synthesised and used as an HTL in OLEDs due to their low ionisation potential and high hole mobility ($\mu_h = 10^{-3}$ cm^2 V^{-1} s^{-1}), see Table 5.4.[23,24] A further advantageous feature of such aromatic diamines is their capability for forming uniform, amorphous layers free of pinholes upon vapour deposition under high vacuum. The glass transition temperature, T_g, of arylamines, such as the arylamine (2) shown in Table 5.1 and those arylamines (12–15)[24,33,34] shown in Table 5.4, is often significantly above room temperature. However, their T_g is often not high enough to avoid posing problems during high-temperature fabrication steps or to completely inhibit crystallisation at moderately high operating temperatures, where crystallisation is accelerated.[33] Nonetheless, they do permit the fabrication of multilayer OLEDs with much longer device lifetimes and a much higher efficiency ($\times 10$) than those of monolayer OLEDs using low-molar-mass electroluminescent materials. For example, bilayer OLEDs utilising compound 15 as the HTL and Alq$_3$ (3) as the ETL and emissive layer have been found to operate at high temperatures ($> 140\,°$C) over many hours without dielectric breakdown.[35] The N,N'-diphenyl-N,N'-*bis*(3-methylphenyl)biphenyl-4,4'-yldiamine (12), often referred to by its acronym (TPD), is frequently used as a nonemissive HTL in multilayer OLEDs.[34] Pyrazoline derivatives, such as compound 16, where the nitrogen atoms are incorporated in an aromatic ring, also exhibit good electron-transport and hole-blocking properties.[36] Spiro-linked versions of the compounds shown in Table 5.4 can also be used to form stable ETL with high glass temperatures.[30,31]

Liquid Crystals as Charge-Carrier Transport Layers

The order present in liquid crystalline phases offers the possibility of attaining the high mobility found in single crystals, but without the traps and defects present at crystal grain boundaries, *i.e.* uniform thin layers with high mobility can be prepared over the area of a display and even lithographically patterned, should this be required. The nature and transport of ions present in nematic liquid crystals had been extensively studied in order to understand their contribution to the electro-optical response of LCDs, especially with regard to the DSM-LCD and TN-LCDs and IPS-LCDs with active matrix addressing, see Chapter 3, well before the use of liquid crystalline phases as transport layers in OLEDs had been proposed.[37–41] However, the charge transport of photogenerated ions (and ionic impurities) through nematic liquid crystals under the action of an applied electric field was also found to be primarily ionic in nature and not electronic.[42–46] This was shown by the viscosity-dependent mobility of ions with a size much larger than electronic charge-carriers in accordance with Walden's rule.[42–46] However, the photogeneration of charge-carriers and subsequent transport under the action of an applied electric field through the ordered columnar liquid crystalline phases and smectic phases has been shown clearly to be electronic.[47–61] Furthermore, the charge-carrier mobility can be high (10^{-1} cm^2 V^{-1} s^{-1} > μ > 10^{-3} cm^2 V^{-1} s^{-1}), almost

Table 5.4 *Molecular structure and glass transition temperature (°C) of the low-molar-mass, non-emissive HTL materials (12–16)*

	Molecular structure	Acronym	T_g	Ref
12		TPD	60	24, 34
13		TTB	82	24
14		NPD	98	24
15		TPTE	130	24
16		PYR-D2	36	

comparable to that through single crystals, for certain compounds in these ordered liquid crystalline phases. Therefore, they may well be suitable as efficient organic transport layers, certainly for holes and perhaps also for electrons, in multilayer OLEDs.

Columnar Liquid Crystals as HTL

Improvements in conductivity could be expected, if the one-dimensionality of the generally amorphous or partially ordered LMMMs used as a HTL could be extended to two dimensions with an increase in order and homogeneity in an organised structure, if crystallisation could be inhibited over the storage and operating temperature range of the device.[60,61] A thin layer of copper phthalocyanine situated between the anode and the HTL is often used in OLEDs in order to reduce the injection barrier of holes into the organic layers. It was recently discovered that triphenylenes[62–67] and phthalocyanines (**17**)[68–70] could exhibit large charge mobility in the columnar (discotic) liquid crystalline state, see Table 5.5 and Figure 5.5. The values observed, *e.g.* in the ordered hexagonal columnar phase of *hexakis*(*n*-alkoxy)triphenylenes (**18**) and in the helical columnar phase of the hexakis(*n*-alkylthio)triphenylenes (**19**)[65] for the charge carrier mobility $(10^{-1}\ cm^2\ V^{-1}\ s^{-1} > \mu_h > 10^{-3}\ cm^2\ V^{-1}\ s^{-1})$ are intermediate between those observed for organic single crystals and inorganic semiconductors $(\mu \approx 1\ cm^2\ V^{-1} s^{-1})$ and those $(10^{-3}\ cm^2\ V^{-1}\ s^{-1} > \mu > 10^{-6}\ cm^2\ V^{-1}\ s^{-1})$ of amorphous conjugated polymers, *c.f.* poly(*N*-vinylcarbazole) (PVK), see Chapter 6. Holes represent the majority charge carrier, rather than electrons, due to the very limited electron affinity and low ionisation potential of the electron-rich aromatic cores of these columnar liquid crystals. The charge transport is non-dispersive and the charge carrier mobility parallel to the director, *i.e.* parallel to the core of the columns of self-assembled discs, is orders of magnitude higher than that measured orthogonal to the director, *i.e.* from column to column. This is partly attributable to the insulating nature of the lateral aliphatic chains surrounding the aromatic cores.

Hence, columnar (discotic) liquid crystals with an aromatic central core[60–82] are of significant interest for application in OLEDs as an HTL, if the columnar phase can be aligned with the columns orthogonal to the substrate surface, *i.e.* the flow of holes from the anode to the ETL and/or emission layer is

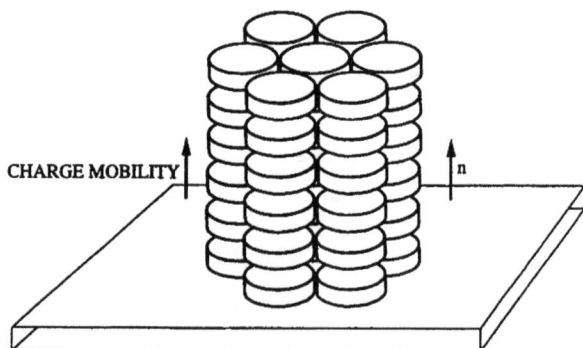

Figure 5.5 *Schematic representation of the* D_{ho} *columnar phase with an ordered hexagonal arrangement of the columns of disc-shaped molecules with a regular period of the discs within the columns. The director is parallel to the columns and normal to the plane of the discs.*

Table 5.5 *Transition temperatures (°C) for the columnar liquid crystals (17–21)*

	Molecular structure	Cr		D_2		D_{ho}		I	Ref
17		● 53		–	●	300		●	70
18 X = O		● 193		–	●	197		●	62–64
19 X = S		● > −5	●	65	●	178		●	60, 61
20		● > −5	●	65	●	178		●	60, 61
21		● 18*	●	83	●	>400		●	98

* D_3–D_2 transition.

facilitated.[60,61] Columnar liquid crystals generally consist of disc-shaped molecules or self-assembled aggregates organised in a supramolecular structure of nearly parallel columns of varying degrees of order in a two-dimensional lattice, see Figure 5.5.[69–73] In the related nematic discotic phase formed by similar or even the same compounds at higher temperatures there is no regular columnar structure.[70–82] In columnar mesophases the molecular cores are organised above each other in columns separated by the peripheral aliphatic chains. Therefore, the intercolumnar distance (≈ 15–$40\,\text{Å}$) is much greater than the intracolumnar distance ($< 4.5\,\text{Å}$), depending to some extent on the length and degree of conformational mobility of the aliphatic chains. The classification depends on the degree and type of order in the lattice (columnar hexagonal, col_h; rectangular, col_r; oblique, col_ob; which are either ordered, *e.g.* col_ho, or disordered, *e.g.* col_hd).[79–82] The nature of columnar phases, especially those with hexagonal symmetry, col_h, in which the aromatic cores of the molecules are arranged above each other in columns allows charge carrier-migration due to the overlap of the π-electron orbitals in the conjugated aromatic core of neighbouring molecules.[83] The ester (**20**) exhibits a plastic columnar phase, which does not crystallise at room temperature.[60,61] The hole mobility in this plastic columnar phase is high ($10^{-2}\,\text{cm}^2\,\text{V}^{-1}\,\text{s}^{-1} > \mu_\text{h} > 10^{-3}\,\text{cm}^2\,\text{V}^{-3}\,\text{s}^{-1}$).

Most compounds exhibiting columnar phases for electrical and electro-optical applications incorporate triphenylene derivatives due to their high tendency to form the required columnar phases and to the presence of the large conjugated aromatic molecular core with a large delocalised π-electron system. In order to be able to synthesise them efficiently in large quantities in high purity and expand the scope for potential columnar liquid crystals with modified structures a series of new synthetic methodologies have had to be developed and optimised.[84–94] They are usually based on the use of palladium-catalysed cross-coupling reactions involving arylboronic acids.[95–97] Such cross-coupling reactions are very tolerant of many functional moieties and enable the synthesis of a wide range of materials not otherwise accessible or with additional protection/deprotection steps.

The largest values for mobility ($\mu \approx 4 \times 10^{-1}\,\text{cm}^2\,\text{V}^{-1}\,\text{s}^{-1}$) have been found[98–100] for columnar liquid crystals with large polyaromatic cores incorporating many phenyl rings, such as the coronene (**21**).[98] However, the very high clearing point of these columnar liquid crystals renders them very difficult to align by cooling from the isotropic liquid into the column phase followed by annealing in the columnar state. This is a common problem for the classes of columnar liquid crystal collated in Table 5.5.

Polycatenar compounds, with more than one terminal chain at each end of a rod-like molecule can also exhibit columnar phases, although they possess a linear molecular structure.[101] Three linear molecules aggregate together to form a self-assembled disc. These discs then stack up to form a fluid columnar structure. Non-dispersive charge-carrier mobility in such polycatenar compounds has also been found.[102]

A further potential problem associated with the practical fabrication of multilayer OLEDs with low-molar-mass calamitic and columnar liquid crystals with a charge-transport layer, especially as the HTL, is the fluid nature of the liquid crystalline state. This could permit contamination of the columnar liquid crystal during the deposition of subsequent layers by the process of vapour deposition or spin-coating. The columnar materials could well be soluble in the solvent used to deposit the next layer. This would lead to chemical contamination of the individual layers due to mixing of one component into another, the formation of layers of uneven thickness and even to the formation of pinholes. Therefore, the Langmuir–Blodgett technique was used to fabricate bilayer and trilayer OLEDs using columnar liquid crystals as the HTL. However, this is not really compatible with low-cost, high-volume device fabrication. Multilayer OLEDs using liquid crystals as the HTL and ETL and/or emission layers created by physical vapour deposition have yet to be reported. Side or main-chain polymers or cross-linked polymer networks could potentially alleviate this fabrication problem, see Chapter 6. However, the polymer structure tends to disrupt the columnar order and lead to significantly lower mobility.[60] Mono-layer OLEDs utilising the ester **20** were found to emit in the blue part of the spectrum (λ_{max} = 500 nm) with a large bandwidth.[60,61]

Smectic Liquid Crystals as HTL and ETL

The compounds (**22–26**) collated in Table 5.6 represent the small number of electroluminescent calamitic liquid crystals.[47–55] Compounds in the liquid crystalline state generally exhibit a higher charge-carrier mobility than in the crystalline state or in the liquid state. This is due to the higher degree of molecular order in the liquid crystalline state without grain-boundary defects and traps associated with polycrystalline compounds and amorphous solids.[47–59] The mobility in the smectic A phase of compound **22** has been shown to be two orders of magnitude higher than that in the isotropic liquid state of the same compound.[47,48] In smectic mesophases the calamic, rod-like molecules are aligned in layers. In the smectic A phase the director is parallel to the molecular long axis and orthogonal to the layer plane, see Figure 5.6. The smectic A phase represents a fluid two-dimensional density wave. The charge-carrier mobility in the ordered lamellar smectic phases of compounds with a rod-like molecular structure, such as compounds (**22–24**) shown in Table 5.6, has been found to be

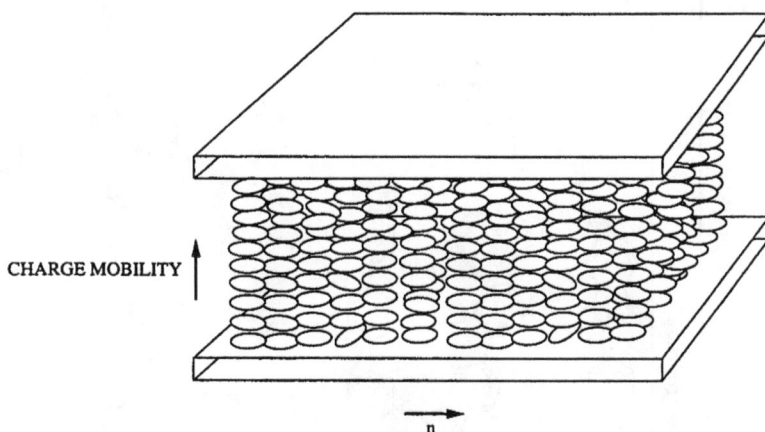

CHARGE MOBILITY

n

Figure 5.6 *Schematic representation of the smectic A phase (SmA) formed by calamitc, rod-like molecules. The director is parallel to the molecular long axis and normal to the plane of the layers.*

Table 5.6 Transition temperatures (°C) for the calamitic liquid crystals (22–26)

Molecular structure	Cr	X	B	A	N	I	Ref
22	•	90 –	–	• 100	–	•	47, 48
23	•	78	–	• 95	–	•	49, 50
24	• 50	• 123*	–	• 128	–	•	51, 52
25	• 48	–	–	–	• 116	•	57, 58
26	• 193	–	–	–	• 197	•	59

* Smectic E phase.

surprisingly high $(10^{-2}$ cm^2 V^{-1} s^{-1} $> \mu > 10^{-4}$ cm^2 V^{-1} s^{-1}) as well as electronic in nature.[47-52] However, the melting point of all three of these compounds is well above room temperature and, therefore, time-of-flight measurements of electric current had to be made at high temperatures. The compounds **22–24** are clearly not suitable for practical applications due to their crystalline nature at room temperature.

The transport of photogenerated current through the smectic A phase of the organic conjugated heterocyclic thiobenzazole (**22**) is nondispersive and primarily of holes ($\mu_h < 5 \times 10^{-3}$ cm^2 V^{-1} s^{-1}) due to the low ionisation potential of this electron-rich compound.[47,48] A high degree of rectification indicated that the majority charge-carrier was holes. The charge-carrier transport through the smectic A phase and the unknown smectic X phase (probably smectic B or smectic E) of the aromatic oxadiazole (**23**) is also unipolar in nature. However, the majority charge-carriers are electrons.[49,50] The mobility in the unknown, but certainly more ordered, smectic X phase ($\mu_e < 0.8 \times 10^{-3}$ cm^2 V^{-1} s^{-1}) is higher than that ($\mu_e = 10^{-4}$ cm^2 V^{-1} s^{-1}) found in the less ordered smectic A phase. The charge-carrier transport through the aromatic phenylnaphthalene (**24**) is ambipolar in nature, transporting holes and electrons almost equally well.[57,58]

Liquid Crystals as Electroluminescent Materials

The ability to macroscopically align the director in the nematic or smectic state formed by rod-like, low-molar-mass liquid crystals on conducting alignment layers offers the possibility of polarised electroluminescent emission.[49,50,53-59] The first report of polarised electroluminescence from a low-molar-mass liquid crystal described emission from an aligned sample of compound **23** in the smectic X phase at temperatures above the recrystallisation temperature well above room temperature.[49,50] Thin layers (15–20 nm) of Alq$_3$ (**3**) and CuPc (copper phthalocyamine) were used as the ETL and HTL, respectively. A thin layer (3 μm) of the compound **23** was the emission layer in a trilayer device, see Figure 5.4. The HTL was deposited by epitaxy on a conductive ITO layer in order to generate the required alignment in the zenithal plane. Polarised emission was observed at 70°C from the smectic X phase at 90 V μm^{-1} and 70 mAcm^{-2}. However, the brightness (0.8 cdm^{-2}) and quantum efficiency (2.8 $\times$ 10^{-5} %) were very low. The polarisation order parameter was also low ($S = 0.32$), probably due to the poor alignment treatment which would give rise to a random pattern of aligned smectic domains.

The nematic compound **25** exhibits a nematic phase above a low melting point.[57,58] However, on quenching this material forms the glassy state with a T$_g$ well above room temperature. Therefore, this material can be uniformly aligned on rubbed polyimide in the relatively fluid nematic state and then quenched by cooling to room temperature in order to fix this order and orientation in the stable glassy state. Nematic glasses of compound **25** have been found to generate polarised photoluminescence and electroluminescence with a high polarisation ratio ($\approx$12:1) and moderate brightness ($L \approx 100$ cdm^{-2}). This material is unique in that it represents the first low-molar-mass, organic electroluminescent material to form a nematic glass above room temperature and then emit plane polarised light with a high polarisation under the action of an applied electric field from the vitrified state. This compound can also be cross-linked to form an insoluble electroluminescent macromolecule, see Chapter 6. Therefore, this compound and related electroluminescent glasses and networks have great potential in combination with a low-cost, clean-up polariser as a source of polarised backlighting for bright LCDs.

Compound **26** exhibits a nematic phase over a narrow temperature range. However, it has been transferred to an OLED substrate by laser ablation under high vacuum without significant photochemical or thermal degradation.[59] If the substrate is an alignment layer, then oriented samples could theoretically be obtained by depositing the electroluminescent material on top of it, which would be another possibility for generating polarised light emission without the need for spin-coating of organic solutions of electroluminescent compounds. This may facilitate the preparation of multilayer OLEDs without the problems associated with layer deposition spin-coating, such as layer corruption and resultant inhomogeneity and non-reproducibility.

Green Electroluminescent, Low-Molar-Mass Organic Materials

Typical examples of low-molar-mass electroluminescent materials, which emit green light are collated in Tables 5.7 and 5.8. The metal chelates (**27–31**) are thermally stable, efficient emitters of green light. They are also capable of being transferred onto a device substrate as uniform thin layers by physical vapour deposition. Most electroluminescent materials incorporate metal atoms from groups II or III from the periodic table. Alq$_3$ (**3**) contains aluminium from group III and, therefore, has a co-ordination number of 6, *i.e.* the aluminium atom requires six covalent or quasi covalent bonds. The nature of the substituent, R, in the aluminium quinolates (**27**)[103,104] can be chosen to modulate the emission maximum depending on the electron-donating or electron-withdrawing effect ($+I$ or $-I$) of the substituent R. Many other metal chelates with quinolinolato ligands have been synthesised.[105,106] However, Alq$_3$ (**3**) is still the most widely used green emitter.[11,12] One reason is the high quantum yield observed for thin layers of pure Alq$_3$ (**3**) in multilayer OLEDs and another is the high T_g. Multilayer OLEDs using Alq$_3$ (**3**) as a combined ETL and emission layer can give rise to very bright displays ($L > 15\,000\,\mathrm{cdm}^{-2}$).

Chelates **28** and **29**[107,108] containing the metal atoms beryllium (BE) and zinc (Zn) from group II in the periodic table have the co-ordination number 4. Therefore, their complexes require two bidentate ligands instead of three for the chelates incorporating metals from group III such as Alq$_3$ (**3**). The metal chelates BeBq$_2$ and Zn(BTZ)$_2$ have been found to be stable electroluminescent materials of high brightness and long life-times, especially as dopant emitters in an HTL.[107] Zn(BTZ)$_2$ (**29**) exhibits a very broad electroluminescent spectrum almost corresponding to white light emission. This would eliminate the problem

Table 5.7 *Emission maximum (nm) and colour for the metal chelates (27–31)*

	Molecular structure	Acronym	λ_{max}	Colour	Ref
27		Alq_3R	530 ± 35	Green	103, 104
28		$BeBq_2$	516	Green	107
29		$Zn(BTZ)_2$	486, 524	Greenish white	108
30		$Tb(acac)_3$	544	Green	110
31		$Tb(acac)_3phen$	543	Green	111

of controlling the light intensity of three individual materials emitting either red, green or blue light in attempts to produce white light emission from an OLED using LMMMs.[12,21,22,109] However, very few materials[24] emit broad-band light covering most of the visible spectrum and at least two emitters are usually required for white light emission. [12,21,22,109] Therefore, white-light emitters are not dealt with as a separate section in this Chapter.

Lanthanide-containing metal chelates such as compounds **30** and **31**[110,111] generally possess sharp emission spectra with narrow band widths. Thus, the terbidium chelates (**30** and **31**) emit almost monochromatic green light. Electroluminescent materials containing lanthanide metal atoms may exhibit very high quantum efficiencies. Terbidium and europium are often used as dopant emitters in inorganic LEDs. The emission results from intramolecular transfer of energy from the excited triplet state to the 4f energy levels of the metal ion. Thus, the normal maximum (25%) for the quantum efficiency due to emission from the singlet state may be overcome by this intramolecular energy transfer. Thus, theoretically, the quantum efficiency of metal chelates containing lanthanides is four times higher than that exhibited for standard electroluminescent organic materials, where only the singlet state decays by radiative emission.

The coumarin (**32**), the coronene (**33**), the triarylamine (**34**), the quinacroline (**35**), the dimethylamino-substituted oxadiazole (**36**) and the naphthalimide (**37**) listed in Table 5.8 are organic electroluminescent materials rather than organometallic chelates. However, all of these materials have been found to exhibit light in the green region of the electromagnetic spectrum, despite a wide variation in molecular structure.[23,24] Coumarin 6 (**32**) is a commercial dye used in some laser applications.[11,12] Coronene **33** is the conjugated polyaromatic core unit of the columnar liquid crystals referred to in Table 5.5.[105] The aromatic, heterocyclic compounds (**34–36**) can also be used as transport layers of either positive or negative charge-carriers.[17,19,112,113] Naphthalimide **37** also emits green light in spite of the small aromatic core of this particular class of organic compound.[113]

The materials collated in Tables 5.7 and 5.8 may be used either as discrete, combined electron-transport-and-emission layers or as electroluminescent dopants at low concentration in a non-emissive, electron-transporting matrix. Many of these materials are self-quenching, so that the luminescent efficiency of doped layers of these materials in an inert matrix, even at very low concentrations ($\approx 0.5\% < c < 5$ %) can exhibit higher quantum efficiencies than those exhibited by thin layers of the pure material itself.

Red Electroluminescent, Low-Molar-Mass Materials

The molecular structures **38–43** shown in Table 5.9 are typical of red, low-molar-mass electroluminescent materials. The number of efficient low-molar-mass electroluminescent materials emitting in the red is relatively limited. The europium complexes (**38** and **39**)[114,115] exhibit narrow emission spectra in the red part of the visible spectrum. The first europium complexes synthesised, such as compound **38**, were not stable enough to be deposited by physical vapour

Table 5.9 *Emission maximum (nm) and colour for the compounds (38–43)*

Molecular structure	Acronym	λ_{max}	Colour	Ref
38	Eu(TTFA)$_3$	620	Red	114
39	Eu(TTFA)$_3$phen	613	Red	115
40	DCM-1	570–620	Red	12
41	Rhodamine metal complex		Orange	24
42	TBPD		Red	24, 105
43	Perylene	600	Orange	105

deposition. However, modified complexes, such as compound (**39**), which could be deposited by this method were subsequently developed.[115] The coumarin dye (**40**) was used at very low concentrations ($\approx 0.5\%$) as a red chromophore with high quantum efficiency ($\Phi = 2 \sim 3\%$) in an ETL of Alq$_3$ (**3**)[12] and or naphthalimide[113] as part of a bilayer OLEDs. The efficiency of these OLEDs with a doped emission layer was found to be higher than that of the corresponding OLEDs with non-doped emission layers. However, even moderately low dopant concentrations ($\approx 2\%$) led to a much lower quantum efficiency due to self-quenching. It was speculated that charge recombination and, therefore, emission takes place on the coumarin molecules. The emission spectrum of the rhodamine metal complexes (41)[24] can be modulated by varying the nature of the counter ion, X. More importantly the thermal stability of these compounds depends crucially on the nature of the counter ion. Only a limited number of modified rhodamine metal complexes (41; X = $GaCl_4^-$, $TaCl_4^-$ and $InCl_4^-$), known originally as red laser dyes, are stable enough to be deposited by physical vapour deposition.[24] The perylene derivative (42),[24,105] and even perylene (43)[105] itself, have been used as red or orange dyes, respectively, in combined electron-transporting and emission layers in multilayer OLEDs.[105] The colour of perylene varies from blue in dilute solution to yellow in the pure crystalline state due to excimer formation at relatively modest concentrations. A shift in the emitted colour or indeed complete quenching of electroluminescence is a problem for many compounds used as efficient chromophores or dye-dopants in OLEDs. Thus, the external quantum efficiency of an OLED with a doped emission layer can be higher for lower concentrations of the dopant. The large *t*-butyl substituents on the perylene derivative **42** serve to inhibit such molecular association by steric hindrance. This, in turn, limits the degree of self-quenching.

Blue Electroluminescent, Low-Molar-Mass Materials

Some typical low-molar-mass, organic electroluminescent materials (**44–49**), which emit in the blue region of the electromagnetic spectrum are listed in Table 5.10. The first blue emitter used in OLEDs was anthracene (**1**) shown in Table 5.1. However, this is a highly crystalline material. Therefore, other organic materials were required, which form the glassy state at room temperature with a high T_g. The oxazoline (**44**) exhibits blue photoluminescence and electroluminescence in the pure state. However, the formation of exiplexes in the presence of a HTL, such as TPD, in a bilayer device gives rise to emission of green rather than blue light.[18] The oxadiazole (**45**) is a good electron-transport material and a highly efficient blue emitter. The non-linear structure tends to inhibit crystallisation.[22] The tetraphenylbutadiene (**46**),[18,23,24,116] the triarylamine (**47**),[18] the pentaphenyl-cyclopentadiene (**48**)[18] and the arylvinylene (**49**)[116] all emit blue light. The triarylamine (**47**)[18] functions well as an efficient hole-transport and blue-emission layer.

The molecular structures (**50–54**) collated in Table 5.11 are typical of blue electroluminescent metal chelates. The azomethin-zinc compound (**50**) exhibits bright blue electroluminescence.[117] The aluminium metal chelates (**51–53**) illustrate attempts to blue-shift the green emission of Alq_3 (**3**) by modifying the number and nature of the ligands.[25] However, the unfilled co-odination sites of the aluminium compound (**52**) render this material intrinsically unstable. The presence of the methyl substituents on compound **52** shields this material from nucleophilic attack. This substantially increases the stability of this compound.

Table 5.10 *Emission maximum (nm) and colour for the aromatic compounds (44–49)*

	Molecular structure	Acronym	λ_{max}	Colour	Ref
44		BBOT	450	Blue	18
45		OXD-4	470–480	Blue	22
46		TPB	430	Blue	18
47		TPA		Blue	18
48		PPCP	465	Blue	18
49		DPVB	440–490	Blue	116

Table 5.11 *Emission maximum (nm) and colour for the compounds (50–54)*

	Molecular structure	Acronym	λ_{max}	Colour	Ref
50		AZM-Zn		Blue	117
51		(QAl)$_2$O	490	Blue	25
52		Q$_2$Al-OAr	511	Blue	25
53		Al(NQ)$_3$	440	Blue	25
54		MgQ$_2$	500	Blue	107

The presence of the electron-withdrawing nitrogen atom at position 5 of the compound (**53**) results in a significant hypsochromic shift to the blue (90 nm) compared to the emission maximum of Alq$_3$ (**3**). The metal chelate (**54**) with the group II magnesium ion emits bright blue light.[25,107]

6 Performance of OLEDs Using LMMMs

The technical data for a commercial prototype OLED using low-molar-mass materials is collated in Table 5.12.[118,119] The brightness and the contrast ratio of a green monochrome OLED with a 256 × 64 pixel dot matrix is comparable with that of a standard commercially available TN-TFT-LCD with active matrix addressing. However, emission of light from an OLED is Lambertian and, therefore, the viewing-angle dependence of contrast of OLEDs is greatly superior to that of most commercially available LCDs. Only LCDs with the optical axis in the plane of the device, such as IPS-TFT-LCDs with active matrix addressing (see Chapter 3) exhibit equally broad viewing angles. Although the power consumption of this OLED panel is higher than that of a comparable LCD with backlighting, the power consumption, external quantum efficiency and luminous efficiency of OLEDs using low-molar-mass materials are comparable or superior to that of inorganic LEDs of comparable size, complexity and colour. Suitably encapsulated and packaged OLEDs exhibit adequate life-times under normal driving conditions. Therefore, the electro-optical performance of OLEDs using LMMMs is clearly sufficient for commercial applications. Indeed, monochrome OLEDs of this type are already commercially available from the Pioneer Corporation for car dashboard applications. Active matrix OLED panels with full-colour and high-information content, being developed jointly by Sanyo and Kodak, are expected in 2001.

Stability of OLEDs Using LMMMs

The temporal stability of OLEDs using low-molar-mass organic compounds

Table 5.12 *Configuration and typical electro-optical performance of a typical bilayer OLED[118] using the low-molar-mass, aromatic diamine TPD (12)[23,34] as HTL and Alq$_3$ (3)[14] doped with a quinacridone dervative (35)[112] as the ETL and emission layer*

Property	Value
Cathode	Al$_{99}$–Li$_1$
Anode	ITO
Hole transport layer (HTL)	(12)
Electron transport layer (ETL) and emission layer	(35)
Threshold voltage, V_{th}	2.5 V
Operating voltage, V_{op}	5.5 V
Quantum efficiency, ϕ	1%
Power conversion efficiency	0.46%
Luminous efficiency	7.4 lm W^{-1}
Current density, I	$10^{-1} - 10^2$ mA cm^{-2}
Colour	Green
Emission maximum, λ_{max}	520 nm
Bandwidth	200 nm
Brightness	300 cd m^{-2}

has been the prime factor over the last 30 years in inhibiting the successful commercialisation of OLEDs using small molecules, *i.e.* the performance of the devices is degraded significantly during the projected life-span of a normal FPD (> 10 000 h). Although most of the organic materials used as an ETL, HTL and emission layer form an amorphous glassy state at room temperature, they still tend to crystallise slowly over time, especially at temperatures above room temperature where the rate of recrystallisation is accelerated.[24,30,120] The small crystals formed initially grow gradually as smaller crystals coalesce into larger crystals. This leads to the formation of discrete grain boundaries between individual crystals. These defects and inhomogenities act as traps and quenching sites and slowly reduce the quantum efficiency of the display. The crystallisation also disrupts the interface between the electrodes and adjacent organic material, where covalent bonding between the electrode metal and the organic material may exist. [33,121] This mechanical stress can even lead to delamination of the electrode surface. The heterojunction in multilayer OLEDs may also be corrupted by the process of crystallisation of one or both of the layers either side of the junction. This can also lead to the diffusion of one material into the other.[122] Since the layers are very thin, this can have severe consequences for the life-time of the display. The electrodes may also degrade over time in the presence of oxygen or moisture giving rise to dark spots due to oxidation.[123] Electrochemical degradation and excited-state chemical reactions of the organic material may also be contributory factors to the degradation of device performance and eventual failure, although their exact nature may be difficult to identify and monitor.[24] These reactions deplete the number of electroactive species. Furthermore, the reaction products then may act as quenching sites, which then reduce the device efficiency even further. An AC drive form instead of a DC driving voltage may also extend the device lifetime.[124] Therefore, it is evident that the stability of OLEDs using low-molar-mass organic compounds is dependent upon a whole range of interrelated factors associated not only with the organic materials used, but also the device configuration, the nature of the electrodes and the device packaging.[33,107,125,126,127]

The use of insoluble, highly cross-linked anisotropic networks created by the polymerisation of photoreactive monomers, eliminates the problem of crystallisation, at least for organic materials, since polymer networks are macromolecular structures incapable of crystallising, see Chapter 6. Furthermore, the fabrication of multilayer devices would be facilitated by the use of a cross-linked stable HTL next to the anode on the solid substrate surface, onto which subsequent layers can be deposited by vapour deposition. Multilayer OLEDs are intrinsically more stable than monolayer devices due to a better balance of charge-carriers and concentration of the charged species away from the electrodes. The synthesis and cross-linking of a suitable aromatic triarylamine derivative with a polymerisable oxetane group at each end of the molecule for use as a HTL has been reported recently.[128]

An additional advantage of the use of reactive, photopolymerisable liquid crystalline monomers with charge transfer or electroluminescent properties is the ability to generate circularly and linearly polarised light.[53-56] This possibility has many potential applications in flat panel display device applications, especially as a source of polarised backlighting for LCDs with improved brightness.[53]

7 References

1 W. Helfrich and W. G. Schneider, *Phys. Rev. Lett.*, 1965, **14**, 229.
2 W. Helfrich and W. G. Schneider, *J. Chem. Phys.*, 1966, **44**, 2902.
3 S. Miyata and H. S. Nalwa, in 'Organic Electroluminescent Materials and Devices' Gordon and Breach, New York, USA, 1997.
4 W. Mehl and W. Bucher, *Z. Phys. Chem.*, 1965, **47**, 76.
5 D. F. Williams and M. Schadt, *Proc. IEEE*, 1970, **58**, 476.
6 J. Dresner, *RCA Rev.*, 1969, **30**, 322.
7 R. E. Kellog, *J. Chem. Phys.*, 1966, **44**, 411.
8 P. S. Vincett, W. A. Barlow, R. A. Hann and G. G. Roberts, *Thin Solid Films*, 1982, **94**, 171.
9 F. J. Campas and M. Goutermann, *Chem. Phys. Lett.*, 1977, **48**, 233.
10 J. Kalinowski, J. Godlewski and Z. Dreger, *Appl. Phys. Lett. A.*, 1985, **37**, 179.
11 C. W. Tang and S. A. Van Slyke, *Appl. Phys. Lett.*, 1987, **51**, 913.
12 C. W. Tang, S. A. Van Slyke and C. H. Chen, *Appl. Phys. Lett.*, 1989, **65**, 3610.
13 M. Abkowitz and D. M. Pai, *Philos. Mag., B.*, 1986, **53**, 193.
14 D. C. Freeman and C. E. White, *J. Am. Chem. Soc.*, 1956, **78**, 2678.
15 M. Era, C. Adachi, T. Tsutsui and S. Saito, *Chem. Phys. Lett.*, 1991, **178**, 488.
16 C. Adachi, S. Tokito, T. Tsutsui and S. Saito, *Jpn. J. Appl. Phys.*, 1988, **27**, L713.
17 C. Adachi, T. Tsutsui and S. Saito, *Appl. Phys. Lett.*, 1989, **55**, 1489.
18 C. Adachi, T. Tsutsui and S. Saito, *Appl. Phys. Lett.*, 1990, **56**, 799 and 1990, **57**, 531.
19 C. Adachi, T. Tsutsui and S. Saito, *Acta Polytech. Scand. Appl. Phys.*, 1990, **170**, 145 and 215.
20 Y. Hamada, C. Adachi, T. Tsutsui and S. Saito, *Jpn. J. Appl. Phys.*, 1992, **31**, 1812.
21 A. Dodabalapur, L. J. Rothberg and T. M. Miller, *Appl. Phys. Lett.*, 1994, **65**, 2308.
22 J. Kido, M. Kimura and K. Nagaim, *Science*, 1995, **67**, 1332.
23 J. Kido, *Bull. Electrochem.*, 1994, **10**, 1.
24 C. H. Chen, J. Shi and C. W. Tang, *Macromol. Symp.*, 1997, **125**, 1.
25 C. H. Chen, J. Shi and C. W. Tang, *Coord. Chem. Rev.*, 1998, **171**, 161.
26 Y. Hamada, C. Adachi, T. Tsutsui and S. Saito, *Optoelectronics*, 1992, **7**, 83.
27 M. Ohta, Y. Sakon, T. Takahashi, C. Adachi and K. Nagai, *USA Pat.*, 5,420,288, 1995.
28 S. Saito, T. Tsutsui , C. Adachi and Y. Hamada, *USA Pat.*, 5,382,477, 1995.
29 J. Kido, C. Ohtaki, K. Hongawa and K. Nagai, *Jpn. J. Appl. Phys. Part 2*, 1993, **32**, L917.
30 J. Salbeck, *Ber. Bunsenges. Phys. Chem.*, 1996, **100**, 1667.
31 J. Salbeck, N. Yu, J. Bauer, F. Weissörtel and H. Bestgen, *Synth. Met.*, 1997, **91**, 209.
32 J. Salbeck, F. Weissörtel and J Bauer, *Macromol. Symp.*, 1997, **125**, 121.
33 J. R. Sheats, H. Antoniadis, M. Hueschen, W. Leonard, J. Miller, R. Moon, D. Roitman and A. Stocking, *Science*, 1996, **273**, 884.
34 M. Stolka, J. F. Yanus and D. M. Pai, *J. Phys. Chem.*, 1984, **88**, 4707.
35 S. Tokito, H. Tanaka, A. Okada and Y. Taga, *Appl. Phys. Lett.*, 1996, **696**, 878.
36 T. Sano, T. Fuji, Y. Nishio, Y. Hamada, K. Shibata and K. Kuroki, *Jpn. J. Appl. Phys.*, 1995, **34**, 3124.
37 G. H. Heilmeier and P. M. Heyman, *Phys. Rev. Lett.*, 1967, **18**, 583.
38 K. Yoshino, S. Hisamatsu and Y. Inuishi, *J. Phys, Soc. Jpn.*, 1972, **32**, 867.

39 T. Yanagisawa, H. Matsumoto and K. Yahagi, *Jpn. J. Appl. Phys.*, 1977, **16**, 45.
40 M. Yamashita and Y. Ameniya, *Jpn. J. Appl. Phys.*, 1978, **17**, 1513.
41 A. Sugimura, N. Matsui, H. Takahashi, H. Sonomura, H. Naito and M. Okuda, *Phys. Rev. B.*, 1991, **43**, 8272.
42 K. Okamoto, S. Nakajima, M. Ueda, A. Itaya and S. Kusabayashi, *Bull. Chem. Soc. Jpn.*, 1983, **56**, 3545.
43 K. Okamoto, S. Nakajima, M. Ueda, A. Itaya and S. Kusabayashi, *Bull. Chem. Soc. Jpn.*, 1983, **56**, 3830.
44 H. Naito, M. Okuda and A. Sugimura, *Phys. Rev. A.*, 1991, **44**, 3434.
45 H. Naito, K. Yoshida, M. Okuda and A. Sugimura, *Jpn. J. Appl. Phys.*, 1993, **73**, 1119.
46 S. Murikami, H. Naito, M. Okuda and A. Sugimura, *J. Appl. Phys.*, 1995, **78**, 4533.
47 M. Funahashi and J.-I. Hanna, *Jpn. J. Appl. Phys.*, 1996, **35**, L703.
48 M. Funahashi and J.-I. Hanna, *Phys. Rev. Lett.*, 1997, **78**, 2184.
49 H. Tokuhisa, M. Era and T. Tsutsui, *Appl. Phys. Lett.*, 1998, **72**, 2639.
50 H. Tokuhisa, M. Era and T. Tsutsui, *Adv. Mater.*, 1998, **8**, 2639.
51 M. Funahashi and J.-I. Hanna, *Appl. Phys. Lett.*, 1997, **71**, 602.
52 M. Funahashi and J.-I. Hanna, *Mol. Cryst. Liq. Cryst.*, 1999, **331**, 516.
53 M. Grell and D. D. C. Bradley, *Adv. Mater.*, 1999, **11**, 895.
54 S. H. Chen, H. Shi, B. M. Conger, J. C. Mastrangelo and T. Tsutsui, *Adv. Mater.*, 1996, **8**, 998.
55 B. M. Conger, H. Shi, S. H. Chen and T. Tsutsui, *Mat. Res. Soc. Symp. Proc.*, 1996, **425**, 239.
56 S. H. Chen, D. Katsis, A. W. Schmidt, J. C. Mastrangelo, T. Tsutsui and T. N. Blanton, *Nature*, 1999, **397**, 506.
57 S. Farrar, A. E. A. Contoret, J. E. Nicholls, M. O'Neill, S. M. Kelly and A. J. Eastwood, *Appl. Phys. Lett.*, 2000, **76**, 2553.
58 A. E. A. Contoret, S. Farrar, L. May, J. E. Nicholls, M. O'Neill, S. M. Kelly and G. J. Richards, *Adv. Mater.*, 2000, **12**, 971.
59 A. E. A. Contoret, S. Farrar, S. M. Kelly, J. E. Nicholls, M. O'Neill and G. J. Richards, *J. Chem. Soc. Chem. Commun.*, in press.
60 G. Lüssem and J. H. Wendorff, *Polym. Adv. Technol.*, 1998, **9**, 443.
61 B. Glüsen, W. Heitz, A. Kettner and J. H. Wendorff, *Liq. Cryst.*, 1996, **20**, 627.
62 D. Adam, D. Haarer, F. Closs, T. Frey, D. Funhoff, K. Siemensmeier, P. Schumacher, and H. Ringsdorf, *Ber. Buns. Phys. Chem.*, 1993, **97**, 1366.
63 D. Adam, F. Closs, T. Frey, D. Funhoff, D. Haarer, P. Schumacher, H. Ringsdorf and K. Siemensmeier, *Phys. Rev. Lett.*, 1993, **70**, 457.
64 C. Catry, M. van der Auweraer, F. C. de Schryver, H. Bengs, K. Häussling, O. Karthaus and H. Ringsdorf, *Makromol. Chem.*, 1993, **194**, 2985.
65 D.,Adam, P. Schumacher, J. Simmerer, K. Häussling, K. Siemensmeier, K. H. Etzbach, H. Ringsdorf and D. Haarer, *Nature*, 1994, **371**, 141.
66 E. O. Arikainen, N. Boden, R. J. Bushby, J. Clemments, B. Movaghar and W. Wood, *J. Mater. Chem.*, 1995, **5**, 2165.
67 J. Simmerer, D. Adam, P. Schumacher, W. Paulus, K. H. Etzbach, K. Siemensmeier, H. Ringsdorf and D. Haarer, *Proceedings of the 26ᵗʰ Arbeitstagung Flüssigkristalle, Freiburg, Germany*, 1996.
68 P. G. Schouten, J. M.Warman, M. P. de Haas, C. F. van Nostrum, G. H. Gelinck, R. J. M. Nolte, M. J. Copyn, J. W. Zwikker, M. K. Engel, M. Hanack, Y. H. Chang and W. T. Ford, *J. Am. Chem. Soc.*, 1994, **116**, 6880.
69 P. G. Schouten, J. M.Warman, M. P. de Haas, J. F. van der Pol and J. W. Zwikker, *J. Am. Chem. Soc.*, 1992, **114**, 9028.
70 B. Blanzat, C. Bartou, N. Tercier, J.-J. André and J. Simon, *J. Am. Chem. Soc.*, 1987, **109**, 6193.
71 S. Chandrasekhar, B. K. Sadashiva, and K. A. Suresh, *Pramana*, 1977, **9**, 471.

72 S. Chandrasekhar, in 'Advances in Liquid Crystals', Acadademic Press, New York, USA, 1982, **Vol. 5**.

73 S. Chandrasekhar, *Mol. Cryst. Liq. Cryst.*, 1993, **14**, 3.

74 J. Billard, J. C. Dubois, C. Vaucher and A. M. Levelut, *Mol. Cryst. Liq. Cryst.*, 1981, **66**, 115.

75 C. Destrade, J. Malthete, H. T. Nguyen and H. Gasparoux, *Phys. Lett.*, 1980, **78A**, 82.

76 R. Fugnitto, H. Strzelecka, J. C. Zann, J. C. Dubois and J. Billard, *J. Chem. Soc. Chem. Commun.*, 1980, 271.

77 J. W. Goodby, P. S. Robinson, B. K.Teo and P. E. Cladis, *Mol. Cryst. Liq. Cryst., Lett.*, 1980, **56**, 303.

78 J. D. Bunning, J. W. Goodby, G. W. Gray and J. B. Lydon, *Proc. Conf. Liq. Cryst. Two-Dimensional Order, Garmisch-Partenkirchen*, 1980, 397.

79 C. Destrade, M. C. Mondon and J. J. Malthete, *J. de Phys. (Paris), Coll. C. 3*, 1979, **40**, 17.

80 C. Destrade, N. H. Tinh, H. Gasparoux, J. Malthete and A. M. Levelut, *Mol. Cryst. Liq. Cryst.*, 1981, **71**, 111.

81 N. H. Tinh, M. C. Mondon-Bernaud, G. Sigaud and C. Destrade, *Mol. Cryst. Liq. Cryst.*, 1981, **65**, 307.

82 C. Destrade, P. Foucher, H. Gasparoux, H. T. Nguyen, A. M. Levelut and J. Malthete, *Mol. Cryst. Liq. Cryst.*, 1984, **106**, 121.

83 C. Mertesdorf, H. Ringsdorf and J. Stumpe, *Liq. Cryst.*, 1991, **9**, 337.

84 I. M. Matheson, O. C. Musgrave and C. J. Webster, *J. Chem. Soc. Chem. Commun.*, 1968, 278.

85 N. Boden, R. C. Borner, R. J. Bushby, A. N. Cammidge and M. V. Jesudason, *Liq. Cryst.*, 1993, **15**, 851.

86 N. Boden, R. J. Bushby and A. N. Cammidge, *J. Chem. Soc. Chem. Commun.*, 1994, 465.

87 R. C. Borner and R. F. W. Jackson, *J. Chem. Soc. Chem. Commun.*, 1994, 845.

88 J. W. Goodby, M. Hird, K. J. Toyne and T., Watson, *J. Chem. Soc. Chem. Commun.*, 1994, 1701.

89 N. Boden, R. J. Bushby, A. N. Cammidge and G. Headdock, *J. Mater. Chem.*, 1995, **5**, 2275.

90 P. Henderson, H. Ringsdorf and P. Schumacher, *Liq. Cryst.*, 1995, **18**, 191.

91 F. Closs, L. Häussling, P. Henderson, H. Ringsdorf and P. Schumacher, *J. Chem. Soc. Perkin Trans. 1*, 1995, 829.

92 P. Henderson, S. Kumar, J. A. Rego, H. Ringsdorf and P. Schumacher, *J. Chem. Soc. Chem. Commun.*, 1995, 1059.

93 N. Boden, R. J. Bushby, A. N. Cammidge and G. Headdock, *Synthesis*, 1995, 31.

94 P. Henderson, H. Ringsdorf and P. Schumacher, *Liq. Cryst.*, 1995, **18**, 191.

95 N. Miyaura T. Yanagi and A. Suzuki, *Synth. Commun.*, 1981, **11**, 513.

96 M. Hird, G. W.,Gray and K. J. Toyne, *Mol. Cryst. Liq. Cryst.*, 1991, **206**, 187.

97 S. J. Cross, M. Hird, A. W. Hall, J. W. Goodby, S. M. Kelly, K. J. Toyne and C. Wu, *Liq. Cryst.*, 1998, **25**, 1.

98 A. M. van de Craats, J. M. Warman, K. Müllen, Y. Geerts and J. D. Brand, *Adv. Mater.*, 1998, **10**, 36.

99 A. M. van de Craats, J. M. Warman, A. Fechtenkötter, J. D. Brand, M. A. Harbison and K. Müllen, *Adv. Mater.*, 1999, **11**, 1469.

100 S. Kumar, D. S. S. Roa and S. K. Prasad, *J. Mater. Chem.*, 1999, **9**, 2751.

101 H.-T. Nguyen, C. Destrade and J. Malthête, *Adv. Mater.*, 1997, **9**, 375.

102 S. Farrar, J. E. Nicholls, M. O'Neill, S. M. Kelly, G. J. Richards and C. Wu, *Liq. Cryst.*, in press.

103 M. Matsumura and T. Akai, *Jpn. J. Appl. Phys.*, 1996, **35**, 5357.

104 P. E. Burrows, Z. Shen, V. Bulovic, D. M. McCarthy, S. R. Forrest, J. A. Cronin and M. E. Thompson, *J. Appl. Phys.*, 1996, **79**, 79991.

105 C. Adachi, S. Tokito, T. Tsutsui and S. Saito, *Jpn. J. Appl. Phys.*, 1988, **27**, L269.
106 Y. Hamada, T. Sano, M. Fujita, T. Fujii, Y. Nishio and K. Shibata, *Jpn. J. Appl. Phys.*, 1993, **32**, L514.
107 Y. Hamada, T. Sano, K. Shibata and K. Kuroki, *Jpn. J. Appl. Phys.*, 1995, **34**, L824.
108 Y. Hamada, T. Sano, H. Fujii, Y. Nishio, H. Takahashi and K. Shibata, *Jpn. J. Appl. Phys.*, 1996, **35**, L1339.
109 M. Strukelj, R. H. Jordan and A. Dodabalapur, *J. Am. Chem. Soc.*, 1996, **118**, 1213.
110 J. Kido K. Nagai and Y. Ohashi, *Chem. Lett.*, 1990, 657.
111 S. Dirr, H.-H. Johannes, J. Schobel, D. Ammermann, A. Bohler, W. Kowalsky and W. Grahn, *SID'97 Digest*, 1997, 778.
112 J. Shi and C. W. Tang, *Appl. Phys. Lett.*, 1997, **70**, 1665.
113 K. Utsugi and S. Takano, *J. Electrochem. Soc.*, 1992, **139**, 3610
114 J. Kido, N. Nagi, Y. Okamoto and T. Skotheim, *Chem. Lett.*, 1991, 1267.
115 R. A. Campos, I. P. Kovalev, Y. Guo, N. Wakali and T. Skotheim, *J. Appl. Phys.*, 1996, **80**, 7144.
116 H. Tokolin, M. Matsura, H. Higashi, C. Hosokawa and T. Kusumoto, *Proc SPIE*, 1993, **38**, 1910.
117 Y. Hamada, T. Sano, M. Fujita, T. Fujii, Y. Nishio and K. Shibata, *Jpn. J. Appl. Phys.*, 1993, **32**, L511.
118 T. Wakimoto, S. Kawami, K. Nagayama, Y. Yonemot, R. Murayama, J. Funaki, H. Sato, N. Nakada and K. Imai, *Asia Display '95*, 1995, 77.
119 T. Wakimoto, R. Murayama, K. Nagayama, Y. Okuda, H. Nakada and T. Thoma, *SID '96 Digest*, 1996, 849.
120 E.-M. Han, L.-M. Do, Y. Niidome and M. Fujihara, *Chem. Lett.*, 1994, 969.
121 L.-M. Do, E.-M. Han, Y. Niidome, M. Fujihara, T. Kanno, S. Yoshida, A. Maeda and A. J. Ikushima, *J. Appl. Phys.*, 1994, **76**, 5118.
122 L.-M. Do, E.-M. Han, Y. Niidome and M. Fujihara, *Mol. Cryst. Liq. Cryst.*, 1995, **267**, 411.
123 P. E. Burrows, V. Bulovic, S. R. Forrest, L. S. Sapochak, D. M. McCarty and M. E. Thompson, *Appl. Phys. Lett.*, 1994, **65**, 2922.
124 S. A. Van Slyke, C. H. Chen and C. W. Tang, *Appl. Phys. Lett.*, 1996, **69**, 2160.
125 C. Adachi, K. Nagi and N. Tomato, *Appl. Phys. Lett.*, 1995, **66**, 2679.
126 Y. Hamada, T. Sano, M. Fujita, Y. Nishio and K. Shibata, *Chem. Lett.*, 1993, 905.
127 H. Antoniadis, M. R. Hueschen, J. N. Miller, R. L. Moon, D. B. Roitman and J. R. Sheats, *Macromol. Symp.*, 1997, **125**, 59.
128 M. S. Bayerl, T. Braig, O. Nuyken, D. C. Müller, M. Groß and K. Meerholz, *Macromol. Rapid Commun.*, 1999, **20**, 224.

Organic Light-Emitting Diodes Using Light-Emitting Polymers

1 Introduction

There are many configurations of electroluminescent display devices[1-14] using very many different organic polymers[15-19], which correspond to the analogous configurations using low-molar-mass materials, see Chapter 5. Single-layer devices demonstrated the principle of LEDs incorporating organic conjugated semiconductor polymers.[1-5] However, bilayer and trilayer devices usually exhibit higher efficiency, lower threshold and operating voltage as well as lower current. This is due to a more balanced charge injection and transport with recombination of the holes and electrons taking place at the interface between layers and not at the interface between the organic material and one of the electrodes. Therefore, the lifetime of multilayer devices tends to be significantly longer than those of comparable single-layer OLEDs.

2 Monolayer Organic Light-Emitting Diodes (OLEDs) Using LEPs

During evaluation of the physical properties of a conjugated organic polymer consisting of a regular alternating pattern of a 1,4-disubstituted phenylene ring and a *trans*-carbon–carbon double (vinylene) bond, *i.e.* poly(*p*-phenylenevinylene) [PPV], compound **1**, see Table 6.1, at the Cavendish Laboratories in Cambridge, UK, it was discovered that this semiconducting polymer emitted yellow-green light, if a sufficiently large electric current passed through it.[1]

Display Configuration

A thin film (≈ 100 nm) of a PPV derivative is sandwiched between two electrodes supported on a solid substrate, such as glass, a flexible plastic substrate, such as poly(propylene), or even a thin film of aluminium, see Figure 6.1. The device substrate provides the mechanical support to the anode, electroluminescent polymer layer and the cathode. The anode is transparent in

179

Table 6.1 *Emission maximum (nm) and colour for the PPV derivatives (1–6)*

Molecular structure	λ_{max}	Colour	Ref
1	550	Green	1
2	626	Red	2–5
3	523	Green	36–40
4	585	Red	41, 42
5	625	Red	43
6	570	Yellow	44

METALLIC
CATHODE

POLYMER
HTL/ETL AND
EMITTER

TRANSPARENT
ANODE

TRANSPARENT
SUBSTRATE

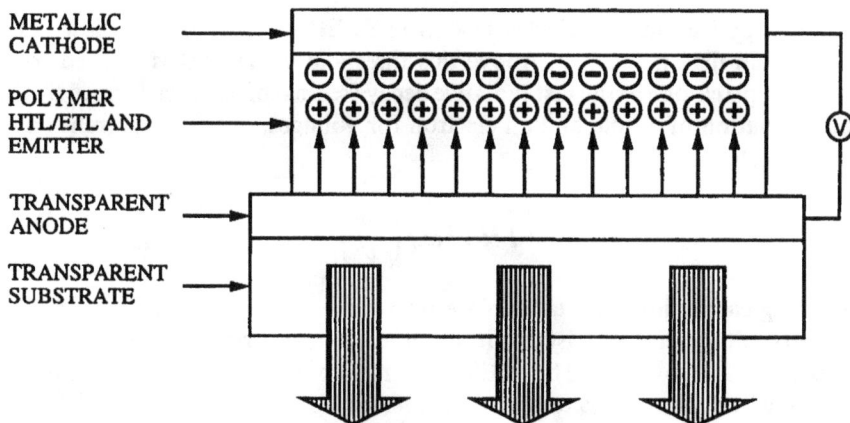

Figure 6.1 *Schematic representation of a monolayer OLED incorporating an electroluminescent conjugated polymer between a transparent anode and a cathode.*

order to allow the light generated by electroluminescence to actually leave the
display and be observed. The electrodes are connected to a source of potential.
Practical displays will also be encapsulated within an inert polymer coating,
especially next to the metallic cathode, in order to prevent the intrusion of
oxygen and moisture into the display. The cathode, *e.g.* a thin layer (≈ 7–15 nm)
of aluminium or calcium, has a low work function, whereas the anode, *e.g.* a
thin metallic layer (15 nm) of indium-tin oxide (ITO) or an organic layer such as
poly(aniline) salt (*e.g.* PANi-camphor sulfonate) or a sulfonic acid derivative of
polyethylenedioxythiophene (PEDOT), has a high work function. No light is
emitted in the absence of an electric current and so the non-activated pixels
appear dark. The application of a voltage between the two electrodes above the
threshold voltage, V_{th}, leads to the injection of electrons from the cathode and
holes from the anode. The high work function of the electrode should match as
closely as possible the highest occupied molecular orbital (HOMO) of the
electroluminescent polymer. Most electroluminescent polymers are organic
semiconductors with a high intrinsic resistivity, *e.g.* 10^{12} Ω cm for PPV (similar
in value to that of typical nematic liquid crystals, which are usually regarded as
dielectrics). Although PPV preferentially transports holes, their mobility is still
low (10^{-4} cm^2 V^{-1} s^{-1} > μ_h > 10^{-8} cm^2 V^{-1} s^{-1}). This means that high
electric fields, *e.g.* 10^5 V cm^{-1}, are required in order to induce an electric current
in the organic layer. However, the polymer layer required for bright, visible
electroluminescence is relatively thin, *e.g.* ≈ 100 nm (even monolayers will emit
easily detectable visible light). Consequently, the threshold voltages and
operating voltages for OLEDs using LEPs are primarily determined by the
thickness of the organic layer, the compatibility of the work functions of the
electrodes and the LUMO and HOMO of the conjugated polymer. Therefore,
threshold voltages and operating voltages for monolayer OLEDs using LEPs
are usually relatively low, *e.g.* 2–10 V for a layer of PPV of standard thickness,
e.g. 100 nm.

Upon application of an electric field to an OLED very little current flows until a threshold field is reached. Above this threshold value light is emitted from activated pixels due to the fast radiative decay (≈ 1 ns) of excitons formed by the recombination of a hole and an electron for voltages, $V > V_{th}$. Current flows according to the relationship:

$$I \propto V^2 \exp\left(\frac{-b}{V}\right) \tag{1}$$

assuming carrier injection takes place by a tunnelling mechanism.[2-5] Different current–voltage characteristics are observed if the current is limited by the mobility of the polymer.[8] The rectification ratio of these diodes, *i.e.* the ratio of current with forward bias to that with reversed bias, is often high, *e.g.* 10^4. The current flowing through an OLED is low, *e.g.* ≈ 10–50 mA cm^{-2}. Thus the operating voltages and power consumption of OLED using LEPs are compatible with use in battery-operated instruments.

The external quantum efficiency, η_{ext}, of monolayer OLEDs, *i.e.* the number of photons actually seen by the observer, is related to the number of photons emitted for every electron injected and is usually low, *e.g.* 0.05% for the first OLEDs using PPV.[1] The observed external efficiency, η_{ext}, is much lower than the internal efficiency, η_{int}, as given by the following relationship:[2-5]

$$\eta_{ext} = \frac{\eta_{int}}{2n^2} \tag{2}$$

where n is the refractive index of the polymer. The external quantum efficiency is substantially lower than the internal quantum efficiency, since the refractive index of conjugated aromatic main-chain polymers is high, *e.g.* $n = 1.4$ for PPV. This suggests that a significant fraction of the light generated is lost within the device itself due to internal reflection and re-absorption leading primarily to non-radiative decay, since, generally, only the singlet excited state relaxes to the ground state with the emission of visible light. The excitons generated by electroluminescence on conjugated polymers can be represented as a quantum mechanical wave, which is localised on one part of the extended polymer chain. The energy of the triplet state is much lower than that of the singlet state and the triplet state usually relaxes to the ground state by a non-radiative pathway. Since symmetry considerations stipulate that three triplets are formed for every singlet, then the maximum luminance efficiency, *i.e.* the percentage of excitons which decay by a radiative pathway, is postulated to be 25%.[8] Quenching at defects and traps can reduce the quantum efficiency significantly. The formation of inter-chain excimers or exciplexes, consisting of one part of a polymer chain in the ground state and one in the excited state, also decay non-radiatively. A substantial fraction of radiative energy is also subsequently lost as heat rather than as visibly emitted light. Therefore, it is not surprising that external quantum efficiencies for OLEDs are intrinsically low. However, acceptable luminance efficiency can still be achieved for monolayer OLEDs, *e.g.* 3 lm W^{-1} for PPV.

Different energy barriers for the injection of holes or electrons exist for an organic polymer with a band gap corresponding to the visible spectrum. This is due to the difference in the work functions (energy required) for the injection of an electron to an organic material (electron affinity) and the injection of a hole, *i.e.* removal of an electron, from the same material (ionisation potential). Therefore, either holes or electrons will preferentially be injected and the charge transport will be unbalanced in favour of one species of charge carrier. This will result in the build up of one charge carrier species at the interface between the organic polymer and either the cathode or anode, where quenching of electroluminescence occurs. Most conjugated organic polymers preferentially transport holes rather than electrons due to their electron-rich aromatic nature, which tends to trap electrons. Therefore, charge capture and recombination with subsequent exciton formation takes place near the cathode, see Figure 6.1. This generally results in a low ($<1\sim2$ %) quantum efficiency, *i.e.* the percentage of photons generated per electron injected, for monolayer OLEDs at low drive voltage ($\approx2\sim3$ V). This may be partly attributable to substantial quenching at the interface between the polymer and the metallic electrode. This interface is not distinct due to a significant degree of covalent or ionic bonding between the metal of the cathode and the organic polymer depending on the nature of both. Indeed this composite metal–polymer interfacial layer between the pure metallic and polymer layers may even lower the energy barrier for charge injection and lead to higher currents at lower voltages. However, in spite of all these possible pathways for loss of observable light, the luminance of monolayer OLEDs may still be comparable with that (100 cd m^{-2}) of a CRT or a large-area, high-information-content LCD, such as a lap-top computer. The latter instruments possess low light-transmission coefficients ($<3\%$) and are sometimes regarded as not particularly bright.[6,11]

The emission spectra of OLEDs are often broad due to the coupling of the exciton energy with the vibrational structure of the polymer in the excited state.[8] The shape of the polymer in the ground (π) state is different from that in the slightly rearranged excited (π^*) state. Therefore, vibrational modes lead to a variation in the effective band gap, which leads to a broadening of the emission spectrum. Therefore, PPV is observed to emit yellow-green light with two maxima ($\lambda_{max} = 520$ and 551 nm).[1] The observed colour of the polymers collated in Tables 6.1–6.16 do not always correspond to λ_{max} due to broad-band emission. Much of the research on improving the properties of electro-luminescent polymers has been devoted to manipulating the band gap between the LUMO and HOMO energy levels in order to determine the colour of light emitted and its bandwidth, see the following section. However, these values are not always quoted, especially in those papers describing the synthesis of new polymers for OLED applications and, therefore, some values are absent from the tables. The absolute values of the LUMO and HOMO are designed to match as closely as possible those of the electrode materials for a particular type of OLED. These can be predicted from the molecular structure using standard theories.[14]

Light-Emitting Polymers (LEPs)

The first reports of electroluminescence from an organic polymer matrix doped with an organic electroluminescent dye in a 'Guest–Host' system failed to attract a great deal of attention.[15,16] However, the report of electroluminescence from poly(p-phenylenevinylene) (PPV) by researchers from the Cavendish laboratory in Cambridge, UK,[1] followed by similar reports using a PPV derivative by researchers from Santa Barbara, USA,[2–4] stimulated enormous interest and activity in organic light-emitting diodes (OLEDs) using light-emitting polymers (LEPs) and the synthesis of new electroluminescent polymers with improved properties.[17–19]

The PPV used in the discovery of electroluminescence was prepared by a soluble precursor route. However, the first synthesis of PPV predates the discovery of its electroluminescent properties by several decades. It was first synthesised using symmetrical phenyl phosphorous Wittig–Horner[20,21] and Wittig reagents[22] and 1,4-phthaldehyde. The resultant polymer was an infusible, intractable solid with a high melting point, which was insoluble in organic solvents. Related PPV derivatives, which were found to be soluble in certain organic solvents, were also prepared using this methodology.[23,24] However, the molecular weight of these polymers, or indeed oligomers, was low on average. Alternative routes to PPV and soluble PPV derivatives using a soluble sulfonium precursor polymer were developed subsequently.[25–27] These methods were then modified[24] in order to produce thin films so that their semiconductor properties could be investigated. Thin transparent films were prepared by depositing a soluble ionic precursor polymer on the desired substrate by spin casting from solution followed by transformation into PPV by heating the film at a high temperature (180°C–300°C) under vacuum for several hours. This procedure has the considerable disadvantage of requiring chemistry to be carried out on the device substrate at high temperatures for a long time. This imposes limitations on the other device components commensurate with these demanding processing conditions. The complete removal of side-products and residual ionic impurities produced during the reaction is not a trivial task. Traces of oxygen lead to the oxidation of parts of the polymer giving rise to carbonyl groups such as aldehyde and acid derivatives. These may act as traps and lead to quenching of electroluminescence, reducing the amount of light emitted.[28–30] Therefore, the processing of the precursor polymer and heat treatment to produce the fully conjugated PPV derivative by elimination are carried out under a dry atmosphere of an inert gas, e.g. nitrogen. However, a series of modified PPV polymers and copolymers with a wide spectrum of physical properties has been successfully produced using this precursor method. Processing temperatures below 160°C[31–35] allow plastic substrates, such as poly(ethylene terephthalate) [PET] or poly(propylene) to be used.[35]

These problems were eliminated by the use of modified PPV derivatives[2–5], such as poly[2-methoxy-5-(2-ethylhexyloxy)-4-phenylene vinylene] (MEH-PPV) compound 2[2], see Table 6.1, which are soluble in organic solvents. The presence of lateral substituents in the polymers 3–6 shown in Table 6.1 induces a lower melting point or glass transition temperature for these polymers compared to that of PPV. Consequently these polymers are sufficiently soluble in organic solvents. This removes the necessity of carrying out chemistry and thermal processing on the substrate. The lateral substituents also represent a means of modulating the LUMO and HOMO levels and, therefore, the wavelength of emission of these PPV derivatives. For example, dialkoxy-substituted PPV derivatives, such as compounds 2–6[2–5,36–44] exhibit a shifted emission compared to that of PPV (1).[1,45–47] Branched chains such as that in MEH-PPV give rise to the largest effects. However, if the substituent is very large, such as in the cholesteryl-substituted PPV derivative (6) then a dilution effect leads to lower quantum efficiencies. Therefore, a compromise must be made between solubility and quantum efficiency.

The ability to control the wavelength of emission and quantum efficiency, ϕ_{eff}, by synthesis is demonstrated by the structures of the modified PPV derivatives 7–11 listed in Table 6.2. The quantum efficiency of polymer 9 is less than half (24%) that (65%) of the similar polymer 11. The presence of one to four additional phenylene rings in the compounds 7–11[49–55] compared to PPV (1) shifts the wavelength of maximum emission by over 130 nm changing the colour from blue-green to red-orange. The presence of the alkyl or alkoxy chain renders most of these materials (7–10) soluble in organic solvents. Compound 11,[55] with two phenylene groups in a lateral position, must be prepared by the precursor route for practical applications, since it is insoluble in organic solvents. However, this lack of solubility in solvents may be of use in the fabrication of multilayer devices, see Section 3.4.

The synthesis of copolymers based on PPV incorporating varying amounts of different structural elements to yield compounds, such as those (12–14)[50,51,56] collated in Table 6.3, allows fine-tuning of their physical properties by choice and number of the structural elements to be combined and the molar ratio of these elements. Once again the presence of long alkyl and alkoxy chains renders high-molecular-weight copolymers, such as 12 and 13, soluble in certain organic solvents. It can be seen that changing the molar ratio (n:m:p) of the three constituent parts of the copolymer (13) allows the colour of emission to be varied from green to yellow to orange. The use of a 1,3-disubstituted phenylene ring in copolymer (13) contributes to the good solubility of this copolymer by maintaining the axis of the lateral alkyl group parallel to the molecular long axis of the copolymer.

The presence of a carbon–carbon triple bond in the poly(phenylene ethynylenes) (15–17) shown in Table 6.4 gives rise to a fully conjugated structure similar to that of the corresponding PPV derivatives with a carbon–carbon double bond instead of the triple bond.[57–61] However, the polymer backbone is somewhat more linear. This results in a blue-shift for poly(phenylene ethynylene) derivatives compared to the corresponding PPVs. Poly(phenylene ethynylenes) such as those (15–17) listed in Table 6.4 are usually prepared by transition metal-catalysed cross-coupling reactions.[57–62] However, despite their ready synthesis they do not appear to offer any appreciable advantages over the corresponding PPV analogues in standard types of OLEDs. However, several of

Table 6.2 *Emission maximum (nm) and colour for the substituted PPV deriva-*
tives (7–11)

Molecular structure	λ_{max}	*Colour*	*Ref*
7	620	Red-orange	49
8	515	Green	50,51
9	560	Green	52
10	510	Blue-green	53
11	490	Blue-green	54

Table 6.3 *Emission maximum (nm) and colour for the copolymer PPV derivatives (12–14)*

	Molecular structure	λ_{max}	Colour	Ref
12		585	Orange Green	50
13			Yellow Orange	51
14			Yellow	56

Table 6.4 *Emission maximum (nm) and colour for the copolymer PPV derivatives (15–17)*

Molecular structure	λ_{max}	Colour	Ref
15	490	Blue	57
16	495	Blue	58–60
17	480	Blue-green	61

these polymers, *e.g.* **15** and **16**, exhibit a thermotropic, enantiotropic nematic phase and have been used to generate polarised light emission, see Section 4.5.

The polymers **18–26** containing the thiophene ring shown in Table 6.5 all exhibit some common features, such as long carbon chains in lateral positions to generate solubility in organic solvents and the 2,5-disubstitution pattern.[63–77] The presence of the different lateral substituents determines their relative and absolute solubility in organic solvents, melting point and glass transition

Table 6.5 *Emission maximum (nm) and colour for the compounds (18–26)*

Molecular structure	λ_{max}	Colour	Ref
18	640	Yellow-orange	63–67
19	640	Yellow-orange	68,69
20	460	Blue	70
21	530	Green	70
22	555	Green	71–73
23	460	Blue	71–73
24	490	Blue	75

(continued)

Table 6.5 *continued*

Molecular structure	λ_{max}	Colour	Ref
25	620	Red	76
26	590, 605	Orange-red	77

temperature. The size of the lateral substituent affects the degree of rotation out of the plane of conjugation of the substituted aromatic rings. The larger and bulkier lateral substituents have the largest effects. A rotation out of the plane of the molecule reduces the degree of conjugation, changes the band gap and, consequently, shifts the wavelength of emission. Thus, the whole of the visible spectrum can be covered by suitably substituted poly(thiophenes).

The conjugated aromatic poly(*p*-phenylene) backbone linked by an aliphatic methylene group in fluorene derivatives constrains the phenylene rings to lie in one plane. This generates a large band gap and, consequently, blue photoluminescent and electroluminescent emission for the poly(fluorenes) (**27–34**) collated in Table 6.6.[78–89] The first blue-emitting OLED using LEPs utilised polymer (**27**) as the emission layer.[78,79] The solubility in organic solvents, melting point and glass transition temperature can all be modulated by the length and nature (branched or unbranched) of the alkyl substituents attached to the fluorene at position 9. Since the substituents at position 9 do not contribute to the degree of conjugation of the poly(*p*-phenylene) backbone the band gap of the polymers (**27–29**) is remarkably uniform. The nature of the monomer components of the copolymers (**29–34**) determines the melting point, glass transition temperature and to some extent the solubility of the resultant copolymer.[83–89] The colour of emission can also be tuned by varying the relative amounts of the monomers in he copolymers, *e.g.* polymer **34**; m = n; λ_{max} = 453 nm.[89]

The poly(*p*-phenylene) derivatives (**35–42**) collated in Table 6.7 all exhibit high band gaps and, consequently, most of them emit in the blue part of the visible spectrum.[75,90–101] Poly(*p*-phenylene) itself (**35**) was first synthesised by a complex precursor route, which generates an insoluble, intractable polymer with a high melting point like PPV (**1**).[90–92] However, subsequent syntheses of poly(phenylene) derivatives has been accomplished much more simply by using transition metal catalysed aryl–aryl cross-coupling reactions.[99–109] The presence of an alkoxy substituent in polymer **36**[93–96] or two alkoxy substituents in polymer **37**[97] renders these polymers soluble in organic solvents. However, the

Table 6.6 *Emission maximum (nm) and colour for the poly(fluorenes) (27–34)*

	Molecular structure	λ_{max}	Colour	Ref
27		470	Blue	78–81
28		477	Blue	82
29		432	Blue	83,84
30		434	Blue	85
31		570	Yellow	86,87
32		590	Orange-red	86,87
33		410	Blue	88

(continued)

Table 6.6 *continued*

	Molecular structure	λ_{max}	Colour	Ref
34		453	Blue	89

presence of the lateral substituents results in rotation of the phenyl ring out of the plane of neighbouring phenyl rings. This interruption of the conjugation results in a further shift in the emission spectrum and a lower quantum yield than that of poly(phenylene). The poly(pyridine) (**38**), with an additional nitrogen atom in the aromatic ring,[98] and the copolymers (**39** and **40**) also emit in the blue.[99,100] The presence of the lateral substituents results in a significant shift due to rotation of the substituted phenylene rings out of the plane of the non-substituted rings. This effect is even more pronounced for the disubstituted poly(phenylenes) (**41** and **42**), where a more significant shift is observed.[101]

The ladder polymers (**43–45**) collated in Table 6.8 are essentially soluble *p*phenylene derivatives. However, the phenyl rings cannot rotate out of the plane of the aromatic core of the polymer due to the presence of the methylene linking unit on the fluorene groups.[110–112] This generally gives rise to a broad emission spectrum. Aggregation of the polymer chain due to the planar nature of the aromatic core unit, *e.g.* for polymer (**43**) can give rise to quenching and time-dependent electroluminescent behaviour.[110] Lateral substituents, such as the methyl groups present in the polymers (**44**), reduce this problem and consistent blue emission with a relatively narrow band width is observed.[111] The presence of random phenyl rings in the stepladder copolymer (**45**) leads to a degree of twisting of the aromatic core sufficient to suppress aggregation, but this is not too prevalent to reduce the quantum efficiency of emission.[112]

It was found that the use of a PPV derivative, *e.g.* compound **46**, see Table 6.9, containing fully conjugated segments separated by non-conjugated linkages, which had not been converted to the conjugated carbon–carbon double bond by elimination of the leaving group from the precursor polymer by annealing actually exhibited a significantly higher quantum efficiency.[113–116] This is a general observation found for a range of conjugated organic polymers exemplified by the typical structures (**46–52**) collected in Table 6.9.[113–131] OLEDs containing partially conjugated PPV were found in some instances to exhibit an external efficiency up to two orders of magnitude higher than that determined for a similar device using fully conjugated PPV.[116] The wavelength of emission (508 nm) is also blue shifted relative to PPV (551 nm) due to the higher band gap of the short conjugated PPV fragments. These polymers illustrate the point that diluting the active chromophore by chemically binding

Table 6.7 *Emission maximum (nm) and colour for the compounds (35–42)*

	Molecular structure	λ_{max}	Colour	Ref
35		459	Blue	90–92
36			Blue	93–96
37		400, 460	Blue	97
38			Blue	98
39		470	Blue	99
40		485	Blue	100
41			Green	101
42			Black	101

Table 6.8 *Emission maximum (nm) and colour for the combined polymers (43–45)*

	Molecular structure	λ_{max}	Colour	Ref
43			Yellow	110
44		420	Blue	111
45		460	Blue	112

Table 6.9 *Emission maximum (nm) and colour for the main-chain polymers (46–52)*

Molecular structure	λ_{max}	Colour	Ref
46	508	Blue-green	113, 114
47	470	Blue	117–119
48	498	Blue-green	120
49	460–550	Blue-green	121–124
50	452	Blue	125–127
51	475	Blue	128
52	447	Blue	129–131

it to a polymer backbone can actually lead to an increase in quantum yield. A short chromophore may maximise efficient emission due to the absence of defects acting as quenching sites often found in a long polymer chain. The absence of exciplex and eximer formation may result in a higher quantum yield per chromophore than that observed for a corresponding fully conjugated polymer. This is counterbalanced to some extent by a lower concentration of emitting parts of the polymer due to the presence of spacers or inert polymer backbone. However, it is difficult to control the degree of conversion and, therefore, the number of non-conjugated linkages and the length of the conjugated PPV segments, since the elimination of leaving groups occurs randomly along the polymer backbone. Furthermore, elimination of some of the linkage leaving groups, *e.g.* sulfonium groups, may occur spontaneously over time in the OLED itself. This will lead to a significant decrease in efficiency over the lifetime of the display.

The problems of reproducibility and stability of partially converted PPV derivatives were solved by the synthesis of a series of main-chain polymers incorporating sections of PPV of defined length separated by inert, non-aromatic, non-conjugated spacer units. The short identical chromophores emit blue light with a narrower bandwidth. This is illustrated by the structures collated in Table 6.9 for some typical main-chain polymers (47–50).[117–131] The transition temperatures and efficiency of emission of the polymers (47–49) may be controlled by varying the relative concentrations of the inert spacer and the PPV segments as well as the length of the segments themselves, *e.g.* n and m for polymer 49.[121–124] It was also found subsequently that a small inert group, such as a silane moiety in polymers 50 and 51,[125–127] served to interrupt the conjugation at well-defined intervals without greatly diluting the active chromophore. The four ethyl groups attached to the silicon atoms, rather than the aromatic rings of polymer 51 containing *p*-quarterterphenyl segments separated by a silane spacer, also act as lateral substituents. Thus, they serve to render this polymer soluble in organic solvents in spite of the high melting point and insolubility of molecular *p*-quarterphenyl.[128] The polymer 52 exhibits an enantiotropic nematic phase.[129–131] This can be made use of for polarised light emission of blue light, see Section 5.

An alternative approach is to dissolve an efficient electroluminescent chromophore in an inert polymer host in a guest–host system or polymer blend.[132–135] Unfortunately, phase separation and demixing often limits the amount of low-molar-mass electroluminescent chromophores that can be dissolved in polymer hosts. Therefore, the relative brightness is lower for such guest–host systems. This problem can be overcome by fixing the chromophore chemically to the polymer itself as a pendent group on a side-chain polymer separated by spacer units. Bulk phase separation is then impossible, although microphase separation may still take place. This is illustrated by the structures collated in Table 6.10 for some typical side-chain polymers 53–55.[136–141]

3   Bilayer OLEDs Using LEPs

There are several possibilities for constructing bilayer OLEDs with a more balanced charge injection. These include an electron-transport layer (ETL) and a combined hole-transport (HTL) and emission layer. Conversely a hole-transport layer and a combined electron-transport and emission layer is also effective.

Table 6.10 *Emission maximum (nm) and colour for the side-chain polymers (53–55)*

Molecular structure	λ_{max}	Colour	Ref
53	620–630	Red	137, 138
54	475	Blue	139
55			141

Display Configuration

A thin film (≈ 100 nm) of a polymer hole-transport layer (HTL) supports a second thin film of a polymer electron-transport layer (ETL) sandwiched between two electrodes supported on a substrate, see Figures 6.2 and 6.3. The anode is transparent in order to allow the passage of the light generated. A potential can be applied between the electrodes. The metal cathode has a low

METALLIC
CATHODE

POLYMER ETL

POLYMER
HTL AND
EMITTER

TRANSPARENT
ANODE

TRANSPARENT
SUBSTRATE

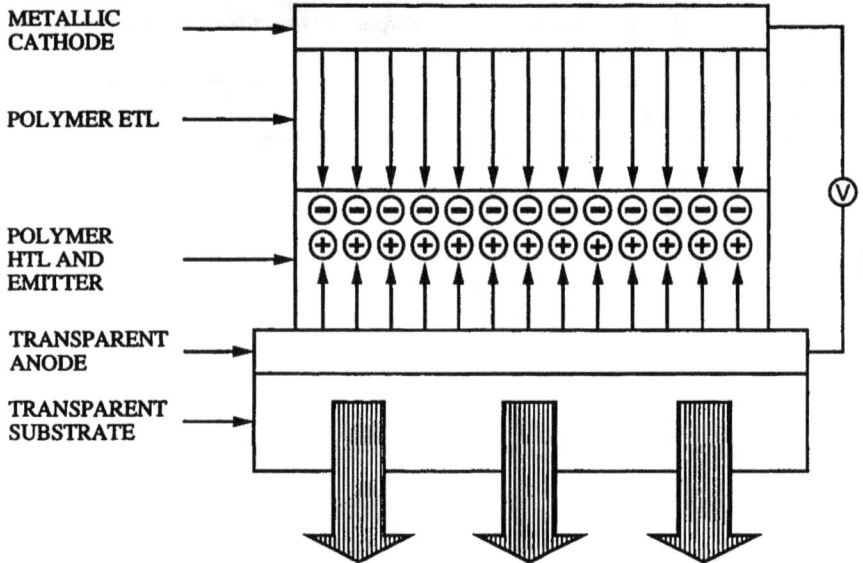

Figure 6.2   *Schematic representation of a bilayer OLED incorporating a combined*
polymer hole-transport and emission layer and a polymer electron-transport
layer situated between a transparent anode and a cathode.

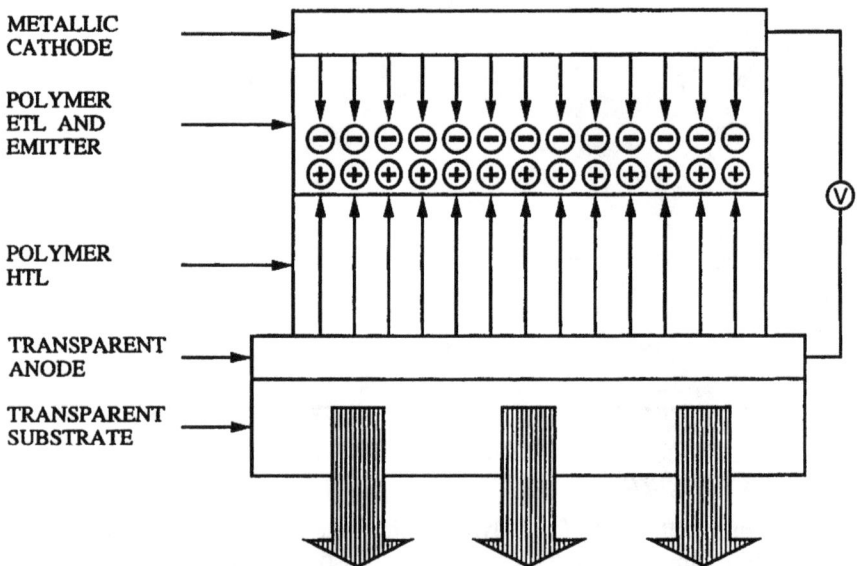

METALLIC
CATHODE

POLYMER
ETL AND
EMITTER

POLYMER
HTL

TRANSPARENT
ANODE

TRANSPARENT
SUBSTRATE

Figure 6.3   *Schematic representation of a bilayer OLED incorporating a polymer hole-*
transport layer and a combined polymer electron-transport and emission layer
situated between a transparent anode and a cathode.

work function, which is matched as closely as possible to the LUMO of the ETL in order to promote the efficient injection of electrons. The HOMO of the HTL is also matched as near as possible to the high work function of the anode, *e.g.* indium–tin oxide (ITO) or an organic polymer electrode such as poly(aniline) salt (PANi). The application of a voltage above the threshold voltage leads to injection of electrons from the cathode into the ETL and holes from the anode into the HTL. The matching of the work functions of the electrode materials to those of the ETL and the HTL minimises the barriers to the injection of holes from the anode and electrons from the cathode. The electrons and holes recombine preferentially at the interface between the ETL and HTL. This results in efficient emission of light from the resultant excitons in one or both of the layers depending on the difference in the band gap of the polymers, see Figures 6.2 and 6.3. Thus, the activated pixels emit light and appear bright.

This configuration generally results in a higher quantum efficiency ($\eta_{ext} < 4\%$) than that observed for comparable monolayer OLEDs due to balanced charge-injection, charge-transport and recombination of holes and electrons away from the electrodes. Very high luminance (≈ 1000 cd m^{-2}) can be achieved at reasonable operating voltages (≈ 6 V).

Polymers as Electron Transport Layers (ETLs)

The use of low-molecular-weight 2,5-disubstituted-1,3,4-oxadiazoles as an ETL in OLEDs using small molecules had been extensively reported, although these often suffer from problems caused by partial crystallisation, see Chapter 5. Therefore, in the first reported bilayer OLED[141] using two polymer layers, a thin layer (30 nm) of 2-(4'-biphenylyl)-5-(4-*t*-butylphenyl)-1,3,4-oxadiazole, PBD (**56**)[142,143], see Table 6.11, was also used as the electron-transport material. However, it was dispersed in an amorphous polymer, *i.e.* poly(methyl methacrylate), in order to inhibit crystallisation. This polymer blend was deposited by spin coating from solution onto an insoluble layer of PPV (**1**). The layer of PPV functioned simultaneously as a hole transport layer and an emission layer. The observed external efficiency of the resultant bilayer device was higher than that of a comparable monolayer device.[144,145] The holes are injected into the PPV layer, travel to the interface between the PPV and PBD/PMMA layer and are blocked by the PBD/PMMA layer at this interface. The field associated with this positive space charge at the layer interface lowers the energy barrier for electron-injection at the cathode. This results in a flow of electrons to the other side of the interface between the PBD/PMMA and PPV layers. The electrons cross over into the PPV layer, combine with holes and the resultant singlet excitons emit visible yellow-green light from the lower band gap PPV layer.

However, the concentration of PDP in the host PMMA matrix had to be low in order to avoid phase separation. This imposes an upper limit on the improvement in external efficiency in using bilayer OLEDs instead of mono-layer OLEDs. Therefore, there was a clear requirement for conjugated organic polymers with a stable, high glass transition temperature and a high affinity for electrons for use as an amorphous, non-crystalline ETL.

Table 6.11 *Molecular structure of the electron-deficient heterocyclic compounds (56–61)*

	Molecular structure	Ref
56		142,143
57		146–149
58		150
59		151, 152
60		153
61		154, 155

Most conjugated aromatic polymers promote the transport of holes rather than electrons. This is a direct consequence of the high density of delocalised electrons attributable to the aromatic rings and carbon–carbon double bonds of a conjugated poly(arylene vinylene). It is much easier to remove an electron than add an electron to an electron-rich organic material. Therefore, most PPV derivatives can function as an HTL or as a combined HTL and emission layer, *e.g.* PPV (1), which preferentially transports holes and blocks the movement of electrons. Polymers for use as electron-transporting and hole-blocking layers in multilayer OLEDs should exhibit a high electron affinity, *i.e.* be electron deficient. Consequently, electronegative atoms and atomic units, such as the strongly electronegative cyano group, were incorporated in modified PPV derivatives, see below, in order to create new strongly polarised, electron-deficient conjugated polymers for use as an ETL.[19] The presence of the electronegative atoms lowers the energy of the LUMO and HOMO levels. This increases the electron affinity of the polymer and promotes electron injection from the cathode.

Main-chain polymers, such as polymer 57 shown in Table 6.11, incorporating a higher concentration the 2,5-disubsituted-1,3,4-oxadiazole moiety directly bound to the polymer backbone were prepared as ETLs.[146–149] The glass transition temperature can be controlled by varying the relative concentration of the monomer units in copolymers.[146–149] Polymers incorporating other electron-deficient heterocyclic rings containing at least one nitrogen atom were found generally to exhibit superior properties.[150–158] Some typical examples are collated in Table 6.11 such as poly(quinoline) (58),[150] poly(1,3,5-trisubstituted triazole) (59),[151,152] poly(1,3,5-trisubstituted triazine) (60)[153] and poly(phenyl-quinoxaline) (61).[154–158] Many of these main-chain polymers also incorporate a number of fluorine atoms attached to an aromatic ring or a methylene spacer unit. The electronegative fluorine atoms contribute to the electron-transporting properties of these polymers. The presence of an oxygen atom between two aromatic rings in the polymers (58–60) gives them a non-linear conformation, which renders them soluble in organic solvents for spincasting. The phenyl rings in a lateral position on polymer 61 serve the same purpose.

Poly(*N*-vinylcarbazole) [PVK] (62), Table 6.12, is a commercial side-chain polymer, which had been found to be a very useful hole-transporting material, although its main application is in photovoltaic devices such as photocopiers. PVK was also used to form an effective HTL as thin uniform film on a device substrate by spincoating, see Table 12.[159] Therefore, an analogous method was used to synthesise sidechain polymers incorporating nitrogen heterocyclic rings as part of pendent groups separated and decoupled from the polymer backbone by inert spacer units for use as ETLs in OLEDs. This allowed the use of standard repeat units, such as acrylates, methacrylates and styrenes, to be used as the polymer backbone. This has many synthetic advantages. A typical example of such a side-chain polymer (63) incorporates the standard (2,5-diphenyl)-1,3,4-oxadiazole ring as the electron-deficient moiety.[160–162] The glass transition temperature of side-chain polymers can be controlled by varying the relative concentration of the monomer units in copolymers such as 64.[163] Side-

Table 6.12 *Molecular structure of some sidechain polymers (62–64)*

	Molecular structure	Ref
62		159
63		160–162
64		163

chain polymers tend to form good quality films. However, the operating voltages of OLEDs using side-chain polymers as transport layers have been found to be high, perhaps due to the dilution of the conjugated aromatic core by the aliphatic spacer and the polymer backbone.

The incorporation of the electron-withdrawing cyano group into the central vinyl linkage between the phenyl rings of PPV derivatives could be relatively easily achieved by the condensation of an aromatic dialdehyde with a diacetonitrile in a Knoevenagel reaction to effectively produce a series of regularly alternating electron-deficient copolymers, such as those (**65–69**) collated in Table 6.13, as an ETL. The electron-donating effects of the lateral alkyl and alkoxy substituents attached to the phenyl rings of the CN-PPV derivatives **65–67** and *para*- and *meta*-substitution patterns result in emission at any part of the visible spectrum.[164–166] The presence of the thiophene ring in the polymer (**68**) results in a small band gap. Thus, it emits in the infrared.[167] The cyano group attached to the carbon–carbon double bond lowers the LUMO level by about 0.6 eV compared to that of corresponding PPV derivatives. There appears to be

Table 6.13 *Emission maximum (nm) and colour for the copolymer PPV derivatives (65–69)*

Molecular structure	λ_{max}	Colour	Ref
65	590	Red	164
66	510	Green	165, 166
67		Blue	165, 166
68	740	IR	167
69			168

no steric effect on the planarity of the polymer attributable to the presence of the bulky cyano group. The flexible nature of the alkyl and alkoxy lateral substituents also confers a good degree of solubility in organic solvents on the polymers so that they can be deposited as thin uniform layers by spincoating. The polymer **69** incorporates an electronegative nitrogen atom in one of the phenyl rings, instead of on the carbon–carbon double bond, in order to give rise to a large value for the electron-affinity of the polymer.[168–171]

Table 6.14 *Emission maximum (nm) and colour for the electron-deficient PPV derivatives (70–74).*

	Molecular structure	λ_{max}	Colour	Ref
70		620–630	Red	63–69
71		620–630	Red	63–69
72		540–570	Yellow-orange	172,173
73		605	red	168
74		630	red	169–171

The PPV derivatives (**70–74**) collated in Table 6.14 incorporate an electron-withdrawing group on the aromatic ring, either as a lateral substituent on the ring or as an heteroatom (N) within the ring, instead of on the vinyl linkage.[63–69,168–173] All of these polymers exhibit a higher electron affinity than unsubstituted PPV (**1**). The band gaps of all of these PPV derivatives are low and emission is in the yellow-orange or red region. However, the high electronegativity of the trifluoromethyl group in the polymer (**72**)[172,173] blue-shifts the emission compared to that of the polymers **70** and **71**[63–69] with either a bromo or chloro substituent. The solubility of the pyridine polymer (**73**)[171,172] is less than that of the corresponding alkyl-substituted pyridine polymer (**74**).[168] The external quantum efficiency of emission of these substituted polymers is not very high and, therefore, they are used purely as an ETL rather than as an emission layer with a high electron affinity.

The polymers (**75–77**) listed in Table 6.15 combine hole-transporting,

Table 6.15 *Emission maximum (nm) and colour for the combined polymers (75–77)*

Molecular structure	λ_{max}	Colour	Ref
75	580	Yellow-orange	175
76	620–630	Blue	149, 163
77	505	Blue	176

electron-transporting and emission functional units in the same polymer.[149,163,174–177] The PPV backbone of polymer **75** acts as an HTL and emission layer at the same time.[175] The oxadiazole side-chain acts as an electron-transport functional group. The polymer **76** combined hole-transporting and electron-transporting functional entities on the same poly(methacrylate) backbone.[149,163] The relative amounts (m and n) of the monomer units determine the final properties of the copolymer. The short PPV fragment results in blue emission. The oxadiazole ring in the polymer **77** assists electron transport through the hole-transporting PPV fragment in the polymer backbone.[176] It also serves to blue-shift the emission. Unfortunately the stability of such combined polymers is sometimes insufficient for practical devices.[149,163]

Bilayer OLEDs using insoluble PPV derivatives as the hole-transporting layer (HTL) are particularly attractive from a fabrication point of view. They can be readily prepared by depositing a soluble precursor and then converting it into an intractable and insoluble film. The second ETL can then be deposited from solution by spin casting. The intractable nature and insolubility of the PPV derivative eliminates problems of dissolution of underlying layers during the deposition and drying of subsequent layers sometimes encountered with soluble polymers. In this case polymers must be chosen which are soluble in one solvent for deposition and insoluble in the solvent solution of the next layer to be deposited. Swelling of polymer layers by solvent absorption can be a serious problem using the technique of polymer deposition by spin-coating from solution.

A bilayer OLED consisting of a layer of the poly(thiophene) (**25**)[76] as ETL next to the cathode and the low-molar-mass PBD (**56**) as HTL[142–144] next to the anode emits white light.[76] Simultaneous emission of red and green light from the poly(thiophene) (**25**) and blue light from PBD (**56**) leads to the mixing of the three primary colours and to the production of white light.

4   Trilayer OLEDs Using LEPs

There are several reports of trilayer OLEDs using three distinct layers for electron transport, hole transport and emission. This device configuration has the advantage that each layer can be optimised for one distinct function, *i.e.* hole-transport, electron-transport and light emission. The problems associated with the fabrication of OLEDs with three distinct polymer layers of controlled thickness, integrity and homogeneity by the technique of deposition from solution by spin-coating are not inconsiderable.

Display Configuration

A thin layer ($\approx$ 100 nm) of a hole-transporting polymer is next to the anode such as ITO. An electron-transporting layer lies next to the cathode. An emission polymer layer is situated between the HTL and the ETL, see Figure 6.4. At least one of the two electrodes is transparent in order to allow the passage of the light generated. An electric field can be applied between the electrodes. The low work function of the metal cathode matches as closely as possible that of the LUMO of the ETL in order to enable efficient electron injection to take place. The high work function of the anode is also matched as closely as possible to the HOMO of the HTL. The electroluminescent polymer in the centre of the display exhibits a high quantum efficiency of emission at the desired wavelength. The application of a voltage above the threshold voltage leads to injection of electrons from

METALLIC
CATHODE

POLYMER
ETL

POLYMER
EMISSION
LAYER

POLYMER
HTL

TRANSPARENT
ANODE

TRANSPARENT
SUBSTRATE

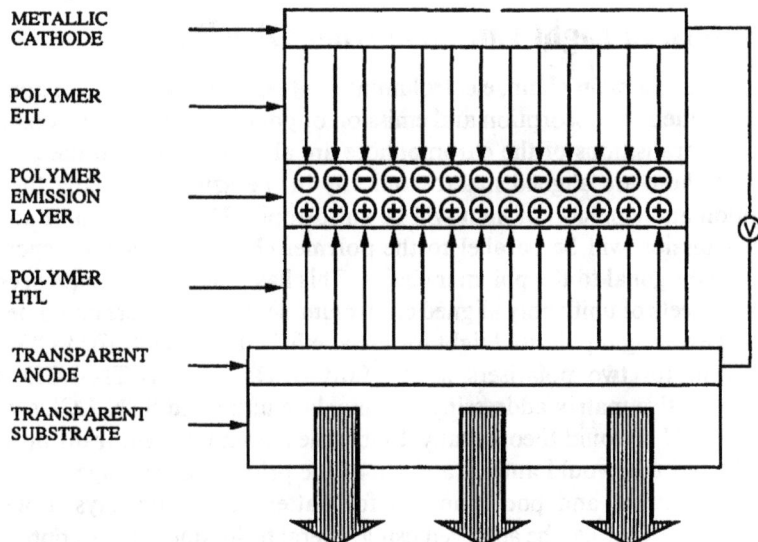

Figure 6.4 *Schematic representation of a trilayer OLED incorporating a polymeric hole-transport layer (HTL), an electron-transport layer (ETL) and a central emission layer situated between a transparent anode and a cathode.*

the cathode into the ETL and holes from the anode into the HTL. The electric current passing through the HTL and ETL polymer layers give rise to a concentration of charge at the interface between them. The electrons and holes then recombine in the emission layer, create excitons and light is emitted.

An early example of a trilayer OLED used PVK (**62**)[159] as the HTL, a polymer blend of PBD (**56**)[142,143] in poly(methacrylate) as the ETL and a poly(quinoxaline) (**61**)[154,155] as the emission layer in the middle of the other two layers.[150] A related device used the same materials as the HTL and ETL, but used segmented PPV derivatives such as those shown in Table 6.9 as the emission layer.[178] The efficiency of these devices may be comparable to that of single-layer OLEDs. However, the operating voltages and current are usually significantly lower, which leads to longer effective device lifetimes.

Conjugated organic polymers such as those shown in the Tables have been used in multilayer OLEDs as the HTL or combined HTL and emission layers or as the ETL or combined ETL and emission layer. The combined polymers (**75–77**) shown in Table 6.15 have been used as combined ETL, HTL and emission layers in various OLED configurations. Blends of these polymers have also been used to maximise OLED efficiency, although phase separation is always a problem with mixtures (blends) of main-chain polymers.[132,133,179–184]

Trilayer OLEDs using LEPs have been fabricated with external quantum efficiencies ($\eta_{eff} \approx 4$–10%) comparable to those obtained for analogous OLEDs using small molecules or LEDs.[140] The low threshold and operating voltages (≈ 5 V) of these multilayer polymer OLEDs means that they can be used as displays in battery operated devices.

5 Polarised Light Emission from OLEDs

If the chains of a main-chain, electroluminescent polymer are aligned parallel to each other then the absorption and emission of polarised light can be observed. The dipole transitions of the chromophore are aligned parallel to the polymer chains. If these are aligned macroscopically in the azimuthal plane then the absorption and emission of light will be anisotropic. The maximum of absorption or emission will be parallel to the polymer chain and the corresponding minima orthogonal to the polymer chains. This has practical consequences, for example, sheets of uniformly aligned electroluminescent polymers could be used as a source of bright polarised light for use as efficient backlights for LCDs. The requirement for two polarisers in standard LCDs, such as TN-LCDs with passive or active matrix addressing and highly multiplexed STN-LCDs would be reduced. This would theoretically double the amount of light transmitted by the LCD and this would ameliorate one of the prime disadvantages of LCDs, *i.e.* low brightness and poor contrast for battery-driven displays. Polarised electroluminescence can be achieved using several techniques under appropriate conditions. However, the potential for polarised electroluminescence was often investigated first by determining the degree of macroscopic alignment of the chromophores by polarised absorption or photoluminescence measurements.

Anisotropic absorption and photoluminescence spectra for polymer blends had been obtained soon after the initial discovery of polymer electroluminescence.[185] Gel-processed polymer blends of MEH-PPV (**2**) in poly(ethylene) were mechanically stretched to uniformly align the long polymer chains parallel to each other in the azimuthal plane. The stretching process unravels and unwinds the coiled polymer chains found in solution and aligns the straightened chains parallel to the direction of stress. This resulted in very high polarisation ratios (60:1) of photoluminescence ($P_{\parallel}/P_{\perp}$) for these blends.[185] More recently photoluminescence, but not electroluminescence, with high polarisation ratios was achieved by essentially the same technique of mechanically stretching polymer blends of poly(arylethynes), such as **15** and **16**,[57–61] in polyethylene.[186] However, the polarisation ratio for electroluminescent devices realised using stretched poly(thiophenes) on a polyethylene substrate was low (2.4:1).[187]

The macroscopic alignment of polymer chains was also achieved by depositing soluble poly(*p*-phenylenes), such as **37**[97] or poly(3-thiophenes) such as **26**,[77] from solution by the Langmuir–Blodgett technique. However, low polarisation ratios ($E_{\parallel}/E_{\perp}$) were found for electroluminescence (3:1).

Low polarisation ratios (< 2:1) for absorption had also been found for amorphous PPV (**1**) deposited from solution by spin-coating on rubbed poly(tetrafluoroethylene) [PTFE].[188] It is evident that this could be improved on by making use of the high order parameter and self-organising properties of the nematic phase of liquid crystalline electroluminescent polymers[189] such as those (**16, 28** and **78–82**) shown in Table 6.16.[16,58–60,81,82,130,190–192] This was then found subsequently to be the case[81,82,86] using thermotropic liquid crystalline polyfluorenes, such as **28** and **80** shown in Table 6.6 and segmented PPV derivatives, such as **81**.[129–131] The nematic phase exhibits the lowest viscosity of

Table 6.16 *Transition temperatures (°C) for the liquid crystalline polymers (16, 28, 78–82)*

Molecular structure	T_g	Cr		LC		I	Ref
78	66	•	103	•	133	•	41
79	100	•	~165	•	~260	•	
80	<93	•	170	•	280	•	81
28	–	•	167	•	>190	•	82
81	77	–		•	179	•	130
16	100	–		•	>135	•	58–60
82	62	–		•	149	•	192

the liquid crystalline phases and, therefore, it is the easiest to macroscopically align by elastic forces on rubbed substrates. Many of the other polymers collated in Tables 6.1–6.15 exhibit thermotropic liquid crystalline phases, although they are often not well characterised. Thermal cross-linking hinders the determination of liquid crystalline transition temperatures of polymers with unsaturated linkages such as poly(*p*-phenylenevinylenes) and poly(arylethynes).

Many of the poly(fluorenes) collated in Table 6.16, the poly(p-phenylenes) collated in Table 6.7 and the segmented PPV derivatives shown in Table 6.9 exhibit an enantiotropic liquid crystalline phase above the melting point. After being aligned in the nematic phase or smectic A phase the macroscopic structure of the homogeneously aligned liquid crystalline phase can be fixed in the glassy state by quenching the polymer. The liquid crystalline phase must be homogeneously aligned on a conducting orientation layer, such as rubbed polyimide with the director parallel to the direction of rubbing. Since polyimide is an insulator, it must be doped with a conducting material[127,128], although some polyimides can also act as HTLs. The best polarised electroluminescence was achieved with an electroluminescence ratio of 15:1 and a switch-on voltage of 18 V from a liquid crystalline poly(fluorene) aligned on rubbed doped polyimide.[189] An electroluminescence polarisation of 12:1 with 200 cd m^{-2} was achieved from PPV (1), which was rubbed when partially formed from its precursor.[189] Alignment of the polymer chains can also be achieved by rubbing the polymer directly on the substrate surface.

The order must then be frozen in before crystallisation occurs, since this would result in the formation of grain boundaries and a reduction in transport or emission efficiency. Device breakdown is also a possibility. The most efficient way to fix the liquid crystalline order is the formation of anisotropic networks by the polymerisation of reactive mesogens in the liquid crystalline state.[189,193–195] Anisotropic polymer networks formed from the thermal or photoinitiated polymerisation of polymerisable, so-called photoreactive, liquid crystalline monomers have been used in a wide variety of electrooptic applications, see Chapter 2.[196–198] This is a more attractive approach than cross-linking aligned sidechain polymers due to the greater ease of orienting low-molar-mass organic materials as a consequence of their much (orders of magnitude) lower viscosity compared to that of polymers.[199] The synthesis of photoreactive oxetanes, deposition on a substrate and subsequent cross-linking to form intractable, insoluble HTL in multilayer OLEDs, has been reported recently, although the monomers were not liquid crystalline.[200]

The first report[193] of this concept did not use a pure photoreactive monomer. However, the photoreactive arylethyne (83),[193] which is not liquid crystalline itself, see Table 6.17, was mixed at low concentration (5 wt%) with the photoreactive liquid crystal 84[196] to form a polymerisable nematic mixture with a high melting point (116°C) and clearing point (150°C). The director of this nematic guest–host mixture, see Chapter 3, was aligned in a cell with rubbed nylon 66 as the alignment layer. This alignment was imposed on the chromophore (83). Photoinitiated polymerisation of the mixture at a high temperature above the melting point of the mixture fixed the mixture in place. A high polarisation ratio ($\approx$11:1) of blue photoluminescence was obtained. Electroluminescence was not reported. The photoreactive monomer (85)[194] exhibits an enantiotropic smectic and nematic phase at high temperature.[194] It was aligned in the nematic state on rubbed polyimide and then polymerised and cross-linked in a thermally initiated polymerisation reaction by raising the temperature of the mixture to 180°C. A moderately high dichroic ratio (3.5) of polarised

Table 6.17 *Transition temperatures (°C) for the photoreactive monomers (83–86)*

Molecular structure	Cr	SmX	N	I	Ref
83	• 80	—	—	•	193
84	• 108	(• 86)*	• 155	•	196
85	• 90	• 144	• 210	•	194
86	• 48	—	• 112	•	195

Parentheses represent a monotropic transition temperature; *SmC absorption was observed. The diene derivative (86)[195] exhibits a low melting point and a moderately high nematic clearing point. No other mesophases could be observed. This allows the nematic director to be aligned on a non-contact alignment layer at moderate temperatures (< 80°C) and then photochemically cross-linked with ultra violet (UV) light using Irgacure (Ciba) as the initiator. The photoalignment layer was doped with 4,4,4-*tris*(naphthyl)-*N*-(phenylamino)triphenylamine, a commercial hole-transporting material, to increase its conductivity. The photoalignment layer was exposed to 0.2 J cm^{-2} of polarised UV light prior to deposition of the photoreactive liquid crystalline diene chromophore (2), which was subsequently cross-linked when in a nematic glassy state by exposure to unpolarised UV light at room temperature using very small amounts of photoinitiator to form a uniform, defect-free cross-linked layer. Using a diene as the photoreactive group has the advantage that it polymerises much more slowly, specifically and selectively than analogous acrylates or methacrylates *via* a sterically controlled intramolecular cyclisation reaction to form a cyclic polymer. The diene derivative (86) with a reactive group at each end of the molecule forms a cross-linked network, see Figures 6.5 and 6.6. Polarised electroluminescent emission of about 100 cd m^{-2} was achieved with a dichroic ratio ($EL_{||}/EL_{\perp}$) of 10:1, see Figure 6.7, and with a threshold voltage of 8 V. This threshold can clearly be improved as a very low threshold voltage of 2 V was achieved when the photoalignment layer was not included in the device, see Figure 6.8. Furthermore, cross-linking does not affect either the photoluminescence or electroluminescence intensity, or the threshold voltage, and actually improves the polarisation ratio of the OLED.

6   Performance of OLEDs Using LEPs

The technical data for a prototype OLED using LEPs are collated in Table 6.18.[201,202] The brightness and the contrast ratio (> 100:1) of monochrome OLEDs with various colours is comparable with that of a standard commercially available TN-LCD with active matrix addressing. The lifetime is clearly

Table 6.18 *Configuration and electro-optical performance of a prototype OLED for backlighting using a soluble PPV derivative as the emission layer*[201,202]

Property	Value
Anode	ITO
Operating voltage, V_{op}	3.5 V
Power consumption	5 mA cm^{-2}
Colour	Green-yellow
Brightness	100 cd m^{-2}
Lifetime (8 cm^2 display)	30 000 h

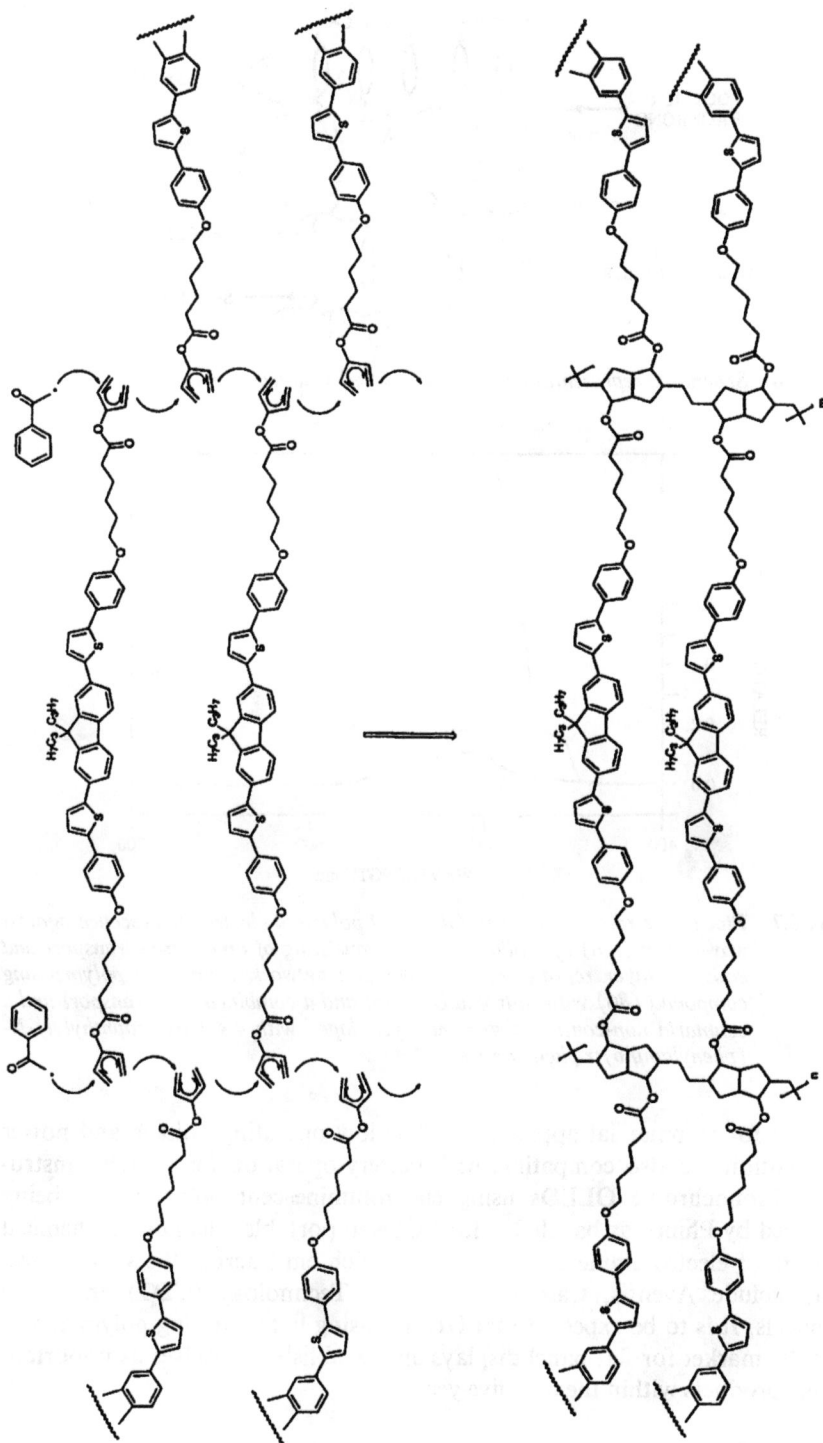

Figure 6.5 *Schematic representation of the polymerisation and crosslinking of compound (86) to form a nematic network.*

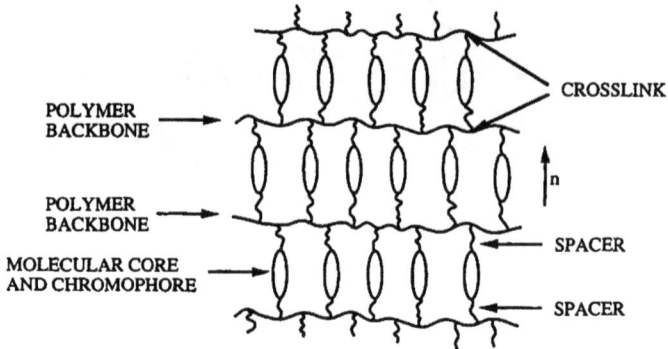

Figure 6.6 *Schematic representation of a cross-linked anisotropic network.*

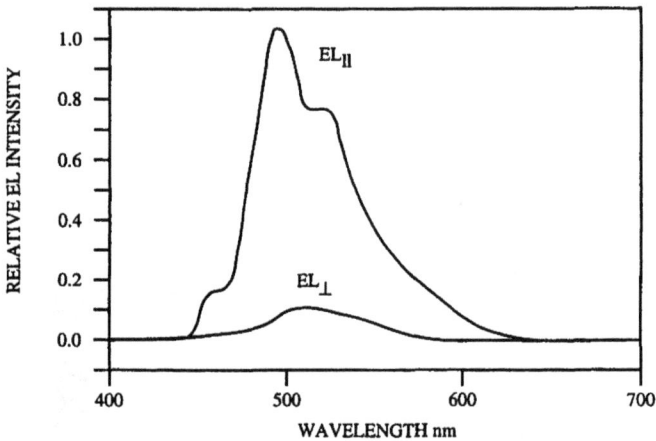

Figure 6.7 *Plot of the relative intensity (A.U.) of polarised electroluminescence against wavelength (nm) of a bilayer OLED consisting of an electron-transport and emission layer represented by a nematic network formed by polymerising compound (86) with isotropic UV light and a combined hole-transport and a coumarin non-contact alignment layer doped with 4,4,4-tris(naphthylyl)-N-(phenylamino)triphenylamine.*

sufficient for commercial applications. The low operating voltage and power consumption are also compatible with battery-operation for portable instruments. Monochrome OLEDs using electroluminescent polymers are being developed by Philips as backlights for LCDs in portable telephones. Chemical suppliers of electroluminescent polymers, which emit across the visible spectrum, include Aventis, Cambridge Display Technology (CDT) and Dow Chemicals. It is to be expected that OLEDs using light-emitting polymers will enter the market for flat-panel displays and establish themselves as important volume products within the next five years.

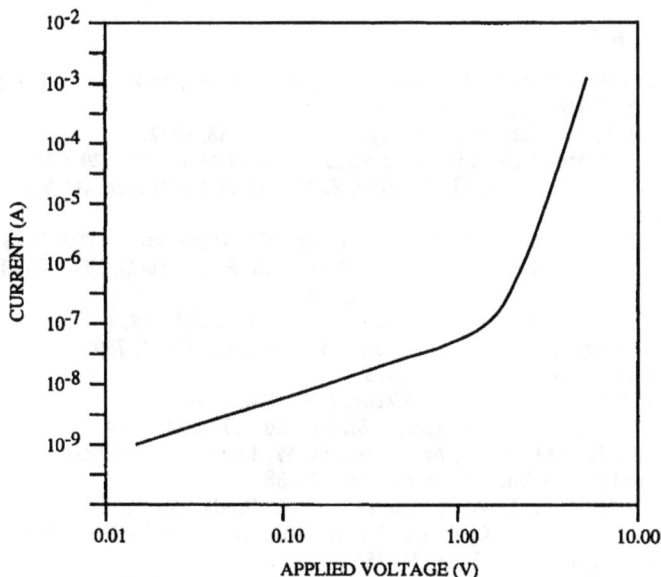

Figure 6.8 *Plot of the current (A) against applied voltage (V) of a bilayer OLED consisting of an electron-transport and emission layer represented by a nematic network formed by polymerising compound (86) with isotropic UV light and a combined hole-transport and a coumarin non-contact alignment layer doped with 4,4,4-tris(naphthylyl)-N-(phenylamino)triphenylamine.*

Stability of OLEDs Using LEPs

The lifetime of simple, alpha-numeric monolayer OLEDs using light-emitting polymers is satisfactory (5000 h < lifetime < 10000 h). The conjugated polymers used in OLEDs form stable amorphous glasses with sufficiently high glass transition temperatures. Therefore, conjugated polymers used in commercial OLEDs do not crystallise over the lifetime of the display. The degradation over time of the electrodes in the presence of oxygen or moisture is due to oxidation and gives rise to visible dark spots in the display.[9–13] However, the major stability problem for OLEDs using light-emitting polymers is the oxidation of carbon–carbon bonds between the aromatic rings of poly(p-phenylene vinylenes).[13] These oxidation reactions reduce the conjugation length of the polymer leading initially to a blue shift in the emission spectrum. Further oxidation can completely deplete the number of electroactive species on a particular polymer chain. In addition, the carbonyl functions generated by the oxidisation act as highly efficient quenching sites. These reduce the external quantum efficiency even further. An AC drive form instead of a DC driving voltage may also extend the device's lifetime. Therefore, the device packaging of OLEDs using electroluminescent polymers is a critical factor in determining the lifetime of OLEDs containing them.[9–13]

7   References

1 J. H. Burroughes, D. D. C. Bradley, A. R. Brown, R. N. Marks, R. H. Friend, P. L. Burn and A. B. Holmes, *Nature*, 1990, **347**, 539.
2 D. Braun and A. J. Heeger, *Appl. Phys. Lett.*, 1991, **58**, 1982.
3 D. Braun, A. J. Heeger and H. Kroemer, *J. Electron Mat.*, 1991, **20**, 945.
4 F. Wudl, P. M. Allemand, G. Srdanov, Z. Ni and D. McBranch, *ACS Symp. Ser.*, 1991, 455.
5 S. Doi, M. Kuwabara, T. Noguchi and T. Ohnishi, *Synth. Met.*, 1993, **55–57**, 4174.
6 R. H. Friend, D. D. C. Bradley and A. B. Holmes, *Phys. World*, 1992, **5**/(11), 42.
7 D. D. C. Bradley, *Synth. Met.*, 1993, **54**, 401.
8 R. H. Friend and N. C. Greenham, *Solid State Phys.*, 1995, **49**, 1.
9 D. D. C. Bradley, *Curr. Opin. Solid State Mater. Sci.*, 1996, **1**, 789.
10 R. W. Gymer, *Endeavor*, 1996, **20**, 115.
11 J. Salbeck, *Ber. Bunsenges. Phys. Chem.*, 1996, **100**, 1667.
12 L. J. Rothberg and A. J. Lovinger, *J. Mater. Res.*, 1996, **11**, 3174.
13 J. R. Sheats, H. Antoniadis, M. Hueschen, W. Leonard, J. Miller, R. Moon, D. Roitman and A. Stocking, *Science*, 1996, **273**, 884.
14 J. Cornil, D. Beljonne, D. A. dos Santos and J. L. Brédas, *Synth. Met.*, 1996, **76**, 101.
15 K. Kaneto, K. Yoshino, K. Koa and Y. Inuishi, *Jpn. J. Appl. Phys.*, 1974, **18**, 1023.
16 R. H. Partridge, *Polymer*, 1983, **24**, 755.
17 A. B. Holmes, D. D. C. Bradley, A. R. Brown, P. L. Burn, J. H. Burroughes, R. H. Friend, N. C. Greenham, R. W. Gymer, D. A. Halliday, R. W. Jackson, A. Kraft, J. H. F. Martens, K. Pichler and I. D. W. Samuel, *Synth. Met.*, 1993, **55–57**, 4031.
18 W. J. Feast, J. Tsibouklis, K. L. Pouwer, L. Groenendaal and E. W. Meijer, *Polymer*, 1996, **37**, 5017.
19 A. Kraft, A. C. Grimsdale and A. B. Holmes, *Angew. Chem. Int. Ed. Eng.*, 1998, **37**, 402.
20 U. Schölkopf, *Angew. Chem.*, 1959, **71**, 260.
21 G. Drehfal, H. H. Hörhold and H. Wildner, *Nachr. Chem. Techn.*, 1965, **13**, 451.
22 R. M. McDonald and T. W. Campbell, *J. Am. Chem. Soc.*, 1960, **82**, 4669.
23 H. H. Hörhold, J. Gottschaldt and J. Opfermann, *J. Prakt. Chem.*, 1977, **319**, 611.
24 H. Martelock, A. Greiner and W. Heitz, *Makromol. Chem.*, 1991, **192**, 967.
25 R. A. Wessling, *J. Polym. Sci., Polym. Symp.*, 1985, **72**, 55.
26 R. W. Lenz, C. C. Han and M. Lux, *Polymer*, 1989, **30**, 1041.
27 T. Momii, S. Tokito, T. Tsutsui and S. Saito, *Chem. Lett.*, 1988, 1201.
28 F. Papadimitrakopoulos, M. Yan, L. J. Rothberg, H. E. Katz, E. A. Chandross and M. E. Galvin, *Mol. Cryst. Liq. Cryst.*, 1994, **256**, 663.
29 F. Papadimitrakopoulos, K. Konstadinidis, T. M. Miller, R. Opila, E. A. Chandross and M. E. Galvin, *Chem. Mater.*, 1994, **6**, 1563.
30 V. H. Tran, V. Massardier, T. P. Nguyen and J. Davenas, *Polymer*, 1996, **37**, 2061.
31 A. Beerden, D. Vanderzande and J. Gelan, *Synth. Met.*, 1992, **52**, 387.
32 R. O. Garay, U. Baier, C. Bubeck and K. Müllen, *Adv. Mater.*, 1993, **5**, 561.
33 M. Herold, J. Gmeiner and M. Schwoerer, *Acta Polym.*, 1994, **45**, 392.
34 M. Herold, J. Gmeiner, W. Riess and M. Schwoerer, *Synth. Met.*, 1996, **76**, 109.
35 G. Gustafsson, Y. Cao, G. M. Treacy, F. Klavetter, N. Colaneri and A. J. Heeger, *Nature*, 1992, **357**, 477.
36 S. Höger, J. J. McNamara, S. Schricker, F. Wudl, *Chem. Mater.*, 1994, **6**, 171.
37 D.-H. Huang, H.-K. Shim, J.-I. Lee and K.-S. Lee, *J. Chem. Soc. Chem. Commun.*, 1994, 2461.
38 C. Zhang, S. Höger, K. Pakbaz, F. Wudl and A. J. Heeger, *J. Electron Mater.*, 1994, **23**, 453.
39 D.-H. Huang, I.-N. Kang, M.-S. Jang, H.-K. Shim and T. Zyung, *Polym. Bull.*, 1996, **36**, 383.

40 S. T. Kim, D.-H. Huang, X.-C. Li, J. Grüner, R. H. Friend, A. B. Holmes and H.-K. Shim, *Adv. Mater.*, 1996, **8**, 979.
41 M. Hamaguchi and K. Yoshino, *Jpn. J. Appl. Phys.*, 1994, **33**, L1478.
42 M. Hamaguchi and K. Yoshino, *Jpn. J. Appl. Phys.*, 1995, **34**, L712.
43 C. Zhang, S. Höger, K. Pakbaz, F. Wudl and A. J. Heeger, *J. Electron Mater.*, 1993, **22**, 413.
44 S. Höger and F. Wudl, *Polym. Preprints*, 1994, **34**, 327.
45 I. Murase, T. Ohnishi, T. Noguchi and M. Hirooka, *Synth. Met.*, 1987, **17**, 639.
46 S. A. Askari, S. D. Rughooputh and F. Wudl, *Synth. Met.*, 1989, **29**, 129.
47 D. A. Haliday, D. D. C. Bradley, P. L. Burn, R. H. Friend and A. B. Holmes, *Synth. Met.*, 1991, **41**, 931.
48 P. L. Burn, D. D. C. Bradley, R. H. Friend, D. A. Halliday, A. B. Holmes, R. W. Jackson and A. Kraft, *J. Chem. Soc. Perkin Trans. 1*, 1992, 3225.
49 S. Tasch, W. Graupner, G. Leising, L. Pu, W.. Wagner and R. H. Grubbs, *Adv. Mater.*, 1995, **7**, 903.
50 Z. Yang, B. Hu and F. E. Karasz, *Macromolecules.*, 1995, **28**, 6151.
51 H. Spreitzer, H. Becker, E. Kluge, W. Kreuder, H. Schenk, R. Demandt and H. Schoo, *Adv. Mater.*, 1998, **10**, 1340.
52 F. H. Boardman, A. W. Grice, M. R. Rüther, T. J. Sheldon, D. D. C. Bradley and P. L. Burn, *Macromolecules*, 1999, **32**, 111.
53 S.-J. Chung, J.-I. Jin and K.-K. Kim, *Adv. Mater.*, 1997, **9**, 551.
54 S.-J. Chung, J.-I. Jin, C.-H. Lee and C.-E. Lee, *Adv. Mater.*, 1998, **10**, 684.
55 B. R. Hsieh, Y. Yu, E. W. Forsythe, G. M. Schaaf and W. A. Feld, *J. Am. Chem. Soc.*, 1998, **120**, 231.
56 H.-P. Weitzel and K. Müllen, *Makromol. Chem.*, 1990, **191**, 2837.
57 A. P. Davey, S. Elliot, O. O'Connor and W. Blau, *J. Chem. Soc., Chem. Commun.*, 1995, 1433.
58 C. Weder, M. S. Wrighton, R. Spreiter, C. Bosshard and P. Günter, *J. Phys. Chem.*, 1996, **100**, 18931.
59 C. Weder, and M. S. Wrighton, *Macromolecules*, 1996, **29**, 5157.
60 D. Steiger, P. Smith and C. Weder, *Macromol. Rapid Commun.*, 1997, **18**, 643.
61 K. Tada, M. Onada, M. Hirohata, T. Kawaii and K. Yoshino, *Jpn. J. Appl. Phys.*, 1996, **35**, L251.
62 R. Giesa, *J. Macromol. Sci. Rev. Macromol. Chem. Phys.*, 1996, **36**, 631.
63 Y. Ohmori, M. Uchida, K. Muro and K. Yoshino, *Jpn. J. Appl. Phys.*, 1991, **30**, L1938.
64 Y. Ohmori, M. Uchida, K. Muro and K. Yoshino, *Solid State Commun.*, 1991, **80**, 605.
65 D. Braun, G. Gustaffson, D. McBranch and A. J. Heeger, *J. Appl. Phys.*, 1992, **72**, 564.
66 M. Uchida, Y. Ohmori, C. Morischima and K. Yoshino, *Synth. Met.*, 1993, **57**, 4168.
67 N. C. Greenham, A. R. Brown, D. D. C. Bradley and R. H. Friend, *Synth. Met.*, 1993, **57**, 4134.
68 A. Bolegnesi, C. Botta, Z. Geng, C. Flores and L. Denti, *Synth. Met.*, 1995, **71**, 2191.
69 M. Pomeranz, H. Yang and Y. Cheng, *Macromolecules*, 1995, **28**, 5706.
70 R. E. Gill, G. G. Malliaras, J. Wildeman and G. Hadziioannou, *Adv. Mater.*, 1994, **6**, 132.
71 M. Berggren, G. Gustafsson, O. Inganäs, M. R. Andersson, O. Wennerström and T. Hjertberg, *Adv. Mater.*, 1994, **6**, 488.
72 M. Berggren, O. Inganäs, G. Gustafsson, J. Ramusson, M. R. Andersson, T. Hjertberg and O. Wennerström, *Nature*, 1994, **372**, 444.
73 O. Inganäs, M. R. Andersson, G. Gustafsson, T. Hjertberg O. Wennerström, P. Dyreklev and M. Grandström, *Synth. Met.*, 1995, **71**, 2121.

74 M. R. Andersson, M. Berggren, O. Inganäs, G. Gustafsson, J. C. Gustafsson-Carlberg, D. Selse, T. Hjertberg and O. Wennerström, *Macromolecules*, 1995, **28**, 7525.

75 N. Tanigaki, H. Masuda and K. Kaeriyama, *Polymer*, 1997, **38**, 1221.

76 M. Berggren, G. Gustaffson, O. Ingenas, M. R. Andersson, T. Hjertberg and O. Wennerström, *J. Appl. Phys.*, 1994, **76**, 7530.

77 A. Bolegnesi, G. Bajo, J. Paloheimo, T. Östergard and H. Stubb, *Adv. Mater.*, 1997, **9**, 121.

78 Y. Ohmori, M. Uchida, K. Muro and K. Yoshino, *Jpn. J. Appl. Phys.*, 1991, **30**, L1941.

79 M. Fukuda, K. Sadawa and K. Yoshino, *Jpn. J. Appl. Phys.*, 1989, **28**, L1433.

80 M. Fukuda, K. Sadawa and K. Yoshino, *J. Polym. Sci. Polym. Chem.*, 1993, **31**, 2465.

81 M. Grell, D. D. C. Bradley, M. Inbasekaran and E. P. Woo, *Adv. Mater.*, 1997, **9**, 798.

82 M. Grell, W. Knoll, D. Lupo, A. Meisel, T. Miteva, D. Neher, H.-G. Nothofer, U. Scherf and A. Yashuda, *Adv. Mater.*, 1999, **11**, 671.

83 H. N. Cho, D. Y. Kim, Y. C. Kim, J. Y. Lee and C. Y. Lee, *Adv. Mater.*, 1997, **9**, 326.

84 H. N. Cho, D. Y. Kim, Y. C. Kim, J. Y. Lee and C. Y. Lee, *Proc SPIE, Int. Soc. Opt. Eng.*, 1997, **3418**, 151.

85 M. Inbasekaran, W. Wu and E. P. Woo, *USA Pat.*, 1998, 5 777 070.

86 M. Grell, M. Redeker, K. S. Whitehead, D. D. C. Bradley, M. Inbasekaran, W. Wu and E. P. Woo, *Liq. Cryst.*, 1999, **26**, 1403.

87 M. Ranger and M. Leclerc, *Mat. Sci. Spec. Iss. Can. I. Chem.*, in press.

88 M. Ranger, D. Rondeau and M. Leclerc, *Macromolecules*, 1997, **30**, 7686.

89 G. Klärner, M. H. Davey, W.-D. Chen, J. Campbell Scott and R. D. Miller, *Adv. Mater.*, 1998, **10**, 993.

90 G. Grem, G. Leditzky, B. Ullrich and G. Leising, *Adv. Mater.*, 1992, **4**, 36.

91 G. Grem, G. Leditzky, B. Ullrich and G. Leising, *Synth. Met.*, 1992, **51**, 383.

92 D. L. Gin and V. P. Conticello, *Trends Polym. Sci.*, 1996, **4**, 217.

93 M. Hamaguchi, H. Sawada, J. Kyokane and K. Yoshino, *Chem. Lett.*, 1996, 527.

94 A.-D. Schlüter and G. Wegner, *Acta Polym.*, 1993, **44**, 59.

95 Y. Yamamoto, *Progr. Polym. Sci.*, 1992, **17**, 1153.

96 Y. Yang, Q. Pei and A. J. Heeger, *J. Appl. Phys.*, 1996, **79**, 934.

97 V. Cimrova, M. Remmers, D. Neher and G. Wegner, *Adv. Mater.*, 1996, **8**, 146.

98 T. Yaamoto, T. Ito and K. Kubota, *Chem. Lett.*, 1988, 153.

99 M. Remmers, M. Schulze and G. Wegner, *Macromol. Rapid Commun.*, 1996, **17**, 239.

100 M. Hamaguchi, H. Sawada, J. Kyokane and K. Yoshino, *Chem. Lett.*, 1996, 527.

101 J. A. John and J. M. Tour, *J. Am. Chem. Soc.*, 1994, **116**, 5011.

102 T. Yamamoto, *Progr. Polym. Sci.*, 1992, **17**, 1153.

103 T. Vahlenkampf and G. Wegner, *Macromol. Chem. Phys.*, 1994, **195**, 1933.

104 T. F. McCarthy, H. Witteler, T. Pakula and G. Wegner, *Macromolecules*, 1995, **28**, 8350.

105 M. Remmers, M. Schulze and G. Wegner, *Macromol. Rapid. Commun.*, 1996, **17**, 239.

106 V. Percec, S. Okita and R. Weiss, *Macromolecules*, 1992, **25**, 1816.

107 V. Percec, J.-Y. Bae, M. Zhao and D. H. Hill, *Macromolecules*, 1995, **28**, 6726.

108 V. Percec, M. Zhao, J.-Y. Bae and D. H. Hill, *Macromolecules*, 1996, **29**, 3727.

109 K. Miyashita and M. Kaneko, *Synth. Met.*, 1995, **68**, 161.

110 U. Scherf and K. Müllen, *Macromol. Chem. Rapid Commun.*, 1991, **12**, 489.

111 S. Tasch, A. Niko, G. Leising and U. Scherf, *Mat. Res. Soc. Symp. Proc.*, 1996, **413**, 71.

112 S. Tasch, A. Niko, G. Leising and U. Scherf, *Appl. Phys. Lett.*, 1996, **68**, 1090.

113 P. L. Burn, A. B. Holmes, A. Kraft, D. D. C. Bradley, A. R. Brown and R. H. Friend, *J. Chem. Soc., Chem. Commun.*, 1992, 32.

114 P. L. Burn, A. B. Holmes, A. Kraft, D. D. C. Bradley, A. R. Brown, R. H. Friend and R. W. Gymer, *Nature*, 1992, **356**, 47.

115 D. Braun, E. G. J. Staring, R. C. J. E. Demandt, G. L. J. Rikken, Y. A. R. R. Kessener and A. H. J. Venhuizen, *Synth. Met.*, 1994, **66**, 75.

116 C. Zhang, D. Braun and A. J. Heeger, *J. Appl. Phys.*, 1993, **73**, 5177.

117 I. Sokolik, Z. Yang, F. E. Karasz and D. C. Morton, *J. Appl. Phys.*, 1993, **74**, 3584.

118 Z. Yang, I. Sokolik and F. E. Karasz, *Macromolecules*, 1993, **26**, 1188.

119 Z. Yang, F. E. Karasz and H. J. Geise, *Macromolecules*, 1993, **26**, 6570.

120· T. Zyung, D.-H. Hwang, I.-N. Kang, H.-K. Shim, W.-Y. Hwang and J.-J. Kim, *Chem. Mater.*, 1995, **7**, 1499.

121 M. Hay and F. L. Klavetter, *J. Am. Chem. Soc.*, 1995, **117**, 7112.

122 G. G. Malliaras, J. K. Herrema, J. Wildeman, R. E. Gill, R. H. Wieringa, S. S. Lampoura and G. Hadzionnou, *Proc. SPIE Int. Soc. Opt. Eng.*, 1993, **2025**, 441.

123 G. G. Malliaras, J. K. Herrema, J. Wildeman, R. H. Wieringa, R. E. Gill, S. S. Lampoura and G. Hadzionnou, *Adv. Mater.*, 1993, **5**, 721.

124 J. K. Herrema, P. F. van Hutten, R. E. Gill, J. Wildeman, R. H. Wieringa and G. Hadzionnou, *Macromolecules*, 1998, **28**, 8102.

125 F. Garten, A. Hilberer, F. Cacialli, E. Esselink, Y. van Dam, B. Schlatmann, R. H. Friend, T. M. Klapwijk and G. Hadziioannou, *Adv. Mater.*, 1997, **9**, 127.

126 H.-J. Brouwer, V. V. Krasnikov, A. Hilberer and G. Hadziioannou, *Adv. Mater.*, 1996, **8**, 935.

127 A. Hilberer, M. Moroni, R. E. Gill, H.-J. Brouwer, V. V. Krasnikov, T.-A. Pham, G. G. Malliaras, S. Veenstra, M. P. L. Werts, P. F. van Hutten and G. Hadzionnou, *Macromol. Symp.*, 1997, **125**, 99.

128 K. Yoshino, K. Tada, M. Hirohata, R. Hidayat, S. Tatsuhara, M. Ozaki, A. Naka and M. Ishikawa, *Jpn. J. Appl. Phys.*, 1997, **36**, L1548.

129 J. Oberski, R. Festag, C. Schmidt, G. Lüssem, J. H. Wendorff, A. Greiner, M. Hopmeier and F. Motamedi, *Macromolecules*, 1995, **28**, 8676.

130 G. Lüssem, R. Festag, A. Greiner, C. Schmidt, C. Unterlechner, W. Heitz, J. H. Wendorff, M. Hopmeier and J. Feldmann, *Adv. Mater.*, 1995, **7**, 923.

131 G. Lüssem, F. Geffarth, A. Greiner, W. Heitz, M. Hopmeier, M. Oberski, C. Unterlechner, and J. H. Wendorff, *Liq. Cryst.*, 1996, **21**, 903.

132 H. Vestweber, J. Oberski, A. Greiner, W. Heitz, R. F. Mahrt and H. Bässler, *Adv. Mater. Opt. El.*, 1993, **2**, 197.

133 H. Vestweber, R Sander, A. Greiner, W. Heitz, R. F. Mahrt and H. Bässler, *Synth. Met.*, 1994, **64**, 141.

134 J. Birgerson, K. Kaeriyama, P. Barta, P. Bröms, M. Fahlman, T. Granlund and W. R. Salaneck, *Adv. Mater.*, 1996, **8**, 982.

135 I.-N. Kang, D.-H. Hwang, H.-K. Shim, T. Zyung and J.-J. Kim, *Macromolecules*, 1996, **29**, 165.

136 J. Pommerehne, H. Vestweber, W. Guss, R. F. Mahrt, H. Bässler, M. Porsch and J. Daub, *Adv. Mater.*, 1995, **7**, 551.

137 X.-C. Li, F. Cacialli, M. Giles, J. Grüner, R. H. Friend, A. B. Holmes, S. C. Moratti and T. M. Young, *Adv. Mater.*, 1995, **76**, 153.

138 F. Cacialli, X.-C. Li, R. H. Friend, S. C. Moriatti and A. B. Holmes, *Synth. Met.*, 1995, **75**, 161.

139 J.-K. Lee, R. R. Schrock, D. R. Baignent and R. H. Friend, *Macromolecules*, 1995, **28**, 1966.

140 D. R. Baignent, N. C. Greenham, J. Grüner, R. N. Marks, R. H. Friend, S. C. Moratti and A. B. Holmes, *Synth. Met.*, 1994, **67**, 3.

141 J. J. M. Halls, D. R. Baignent, F. Cacialli, N. C. Greenham, R. H. Friend, S. C. Moratti and A. B. Holmes, *Thin Solid Films*, 1996, **276**, 13.

142 J.-L. Bredas and A. J. Heeger, *Chem. Phys. Lett.*, 1994, **217**, 507.

143 C. W. Tang and S. A. Van Slyke, *Appl. Phys. Lett.*, 1987, **51**, 913.
144 C. Adaichi, S. Tokito, T. Tsutsui and S. Saito, *Jpn. J. Appl. Phys.*, 1998, **28**, L269.
145 A. R. Brown, D. D. C. Bradley, J. H. Burroughes, R. H. Friend, N. C. Greenham, P. L. Burn, A. B. Holmes and A. Kraft, *Appl. Phys. Lett.*, 1992, **61**, 2793.
146 Y. Yang and Q. Pei, *J. Appl. Phys.*, 1995, **77**, 4807.
147 E. Buchwald, M. Meier, S. Karg, P. Pösch, H.-W. Schmitt, P. Strohriegel, W. Riess and M. Schwoerer, *Adv. Mater.*, 1995, **7**, 839.
148 Q. Pei and Y. Yang, *Chem. Mater.*, 1995, **7**, 1568.
149 X.-C. Li, A. Kraft, R. Cervini, G. C. W. Spencer, F. Cacialli, R. H. Friend, J. Grüner, A. B. Holmes, J. C. DeMello and S. C. Moratti, *Mat. Res., Symp. Proc.*, 1996, **413**, 13.
150 I. D. Parker, Q. Pei and M. Marrocco, *Appl. Phys. Lett.*, 1994, **65**, 1272.
151 M. Strukelj, F. Papadimitrakopoulos, T. M. Miller, L. J. Rothberg, *Science*, 1995, **267**, 1969.
152 M. Strukelj, T. M. Miller, F. Papadimitrakopoulos and S. Son, *J. Am. Chem. Soc.*, 1995, **117**, 11976.
153 R. Fink, C. Frenz, M. Thelekkat and H.-W. Schmitt, *Macromol. Symp.*, 1997, **125**, 151.
154 D. O'Brien, M. S. Weaver, D. G. Lidzey and D. D. C. Bradley, *Appl. Phys. Lett.*, 1996, **69**, 881.
155 M. Thelekkat, R. Fink, P. Pösch, J. Ring and H.-W. Schmitt, *Polym. Prepr.*, 1997, **38**, 323.
156 T. Kanabara and T. Yamamoto, *Chem. Lett.*, 1993, 1459.
157 T. Yamamoto, T. Inoue and T. Kanabara, *Jpn. J. Appl. Phys.*, 1994, **33**, L250.
158 T. Fukada, T. Kanabara, T. Yamamoto, K. Ishikawa, H. Takezoe and A. Fukuda, *Appl. Phys. Lett.*, 1996, **68**, 2346.
159 T. Fujii, M. Fujita, Y. Hamada, K. Shibata, Y. Tsujino and K. Kuroki, *J. Photopolym. Sci. Technol.*, 1991, **4**, 135.
160 M. Meier, E. Buchwald, S. Karg, P. Pösch, M. Greczmiel, P. Strohriegel and W. Riess, *Synth. Met.*, 1996, **76**, 95.
161 M. Greczmiel, P. Pösch, H.-W. Schmitt, P. Strohriegel, E. Buchwald, M. Meier, W. Riess and M. Schwoerer, *Makromol. Symp.*, 1996, **102**, 371.
162 X.-C. Li, T. M. Yong, J. Grüner, A. B. Holmes, S. C. Moratti, F. Cacialli and R. H. Friend, *Synth. Met.*, 1997, **84**, 437.
163 X.-C. Li, F. Cacialli, M. Giles, J. Grüner, R. H. Friend, A. B. Holmes, S. C. Moratti and T. M. Yong, *Adv. Mater.*, 1995, **7**, 898.
164 N. C. Greenham, S. C. Moratti, D. D. C. Bradley, R. H. Friend and A. B. Holmes, *Nature*, 1993, **365**, 628.
165 E. G. J. Staring, R. C. J. E. Demandt, D. Braun, G. L. J. Rikken, Y. A. R. R. Kessener, A. H. J. Verhuizen, M. M. F. van Knippenberg and M. Boumans, *Synth. Met.*, 1995, **71**, 2179.
166 S. C. Moratti, R. Cervini, A. B. Holmes, D. R. Baigent, R. H. Friend, N. C. Greenham, J. Grüner and P. J. Hamer, *Synth. Met.*, 1995, **71**, 2117.
167 D. R. Baigent, P. J. Hamer, R. H. Friend, S. C. Moratti and A. B. Holmes, *Synth. Met.*, 1995, **71**, 2175.
168 M. J. Marsella, D.-K. Fu and T. M. Swager, *Adv. Mater.*, 1995, **7**, 145.
169 M. J. Marsella and T. M. Swager, *Polym. Prepr.*, 1992, **33**, 1196.
170 J. Tian, C.-C. Wu, M. E. Thompson, J. C. Sturm, R. A. Register, M. J. Marsella and T. M. Swager, *Adv. Mater.*, 1995, **7**, 395.
171 J. Tian, C.-C. Wu, M. E. Thompson, J. C. Sturm and R. A. Register, *Chem. Mater.*, 1995, **7**, 2190.
172 A. Lux, A. B. Holmes, R. Cervini, J. E. Davies, S. C. Moratti, J. Grüner, F. Cacialli and R. H. Friend, *Synth. Met.*, 1997, **84**, 293.
173 A. C. Grimsdale, X.-C. Li, F. Cacialli, J. Grüner, A. B. Holmes, S. C. Moratti and R. H. Friend, *Synth. Met.*, 1996, **76**, 165.

174 Z. Boa, Z. Peng, M. E. Galvin and E. A. Chandross, *Chem. Mater.*, 1998, **10**, 1201.
175 Z. Peng and M. E. Galvin, *Chem. Mater.*, 1998, **10**, 1785.
176 Z. Peng, Z. Boa and M. E. Galvin, *Chem. Mater.*, 1998, **10**, 2086.
177 G. Lüssem and J. H. Wendorff, *Polym. Adv. Technol.*, 1998, **9**, 443.
178 M. Remmers, D. Neher, J. Grüner, R. H. Friend, G. H. Gelinck, J. M. Warman, C. Quattrocchi, D. A. dos Santos and J.-L. Brédas, *Macromolecules*, 1996, **29**, 7432.
179 G. Yu, H. Nishino, A. J. Heeger, T.-A. Chen and R. D. Rieke, *Synth. Met.*, 1995, **72**, 249.
180 J. Birgerson, M. Fahlman, P. Bröms and W. R. Saleneck, *Synth. Met.*, 1996, **80**, 125.
181 L.-N. Kang, D.-H. Hwang, H.-K. Shim, T. Zyung and J.-J. Kim, *Macromolecules*, 1996, **29**, 165.
182 N. Tanigaki, H. Masuda and K. Kaeriyama, *Polymer*, 1997, **38**, 1221.
183 H. Westweber, A. Greiner, U. Lemmer, R. F. Mahrt, R. Richert, W. Heitz and H. Bässler, *Adv. Mater.*, 1992, **4**, 661.
184 T. W. Hagler, K. Pakbaz, K. F. Voss and A. J. Heeger, *Phys. Rev.*, 1991, **44**, 8652.
185 C. Weder, C. Sarwa, C. Bastiaansen and P. Smith, *Adv. Mater.*, 1997, **9**, 1035.
186 C. Weder, C. Sarwa, A. Montali, C. Bastiaansen and P. Smith, *Science*, 1998, **279**, 837.
187 P. Dyreklev, M. Berggren, O. Ingenas, M. R. Andersson, O. Wennerström and T. Hjertberg, *Adv. Mater.*, 1995, **7**, 43.
188 K. Pichler, R. H. Friend, P. L. Burn and A. B. Holmes, *Synth. Met.*, 1993, **55–57**, 454.
189 M. Grell and D. D. C. Bradley, *Adv. Mater.*, 1999, **11**, 895.
190 Z. Bao, Y. Chen, R. Cai and L. Yu, *Macromolecules*, 1993, **26**, 5281.
191 Z. Bao, Y. Chen and R. Cai, *Angew. Chem. Int. Ed. Eng.*, 1993, **32**, 1345.
192 Y. Watanabe, T. Mihara and N. Koide, *Macromol. Chem. Phys.*, 1998, **199**, 977.
193 A. P. Davey, R. G. Howard and W. J. Blau, *J. Mater. Chem.*, 1997, **7**, 417.
194 A. Bacher, P. G. Bentley, D. D. C. Bradley, L. K. Douglas, P. A. Glarvey, M. Grell, K. S. Whitehead and M. L. Turner, *J. Mater. Chem.*, 1999, **9**, 2985.
195 A. E. A. Contoret, S. R. Farrar, P. O. Jackson, L. May, M. O'Neill, J. E. Nicholls, S. M. Kelly, and G. J. Richards, *Adv. Mater.*, 2000, **12**, 971.
196 D. J. Broer, *Adv. Polym. Sci.*, 1994, **118**, 1.
197 S. M. Kelly, *J. Mater. Chem.*, 1995, **5**, 2047.
198 S. M. Kelly, *Liq. Cryst.*, 1998, **24**, 71.
199 X.-C. Li, T.-M. Yong, J. Grüner, A. B. Holmes, S. C. Moratti, F. Cacialli and R. H. Friend, *Synth. Met.*, 1997, **84**, 437.
200 M. S. Bayerl, T. Braig, O. Nuyken, D. C. Müller, M. Groß and K. Meerholz, *Makromol. Chem. Rapid Commun.*, 1999, **20**, 224.
201 Philips Web Site.
202 CDT Web Site and the Cavendish Laboratory Web Site.

CHAPTER 7

Conclusions and Outlook

The various types of liquid crystal displays (LCDs) undeniably represent the dominant flat panel display technology at the present time and probably for the foreseeable future.[1-5] In many cases they have not replaced a display unit based on another technology in an existing instrument or device, but have provided the enabling technology to allow a new type of instrument requiring a visual display unit to be manufactured or rendered it commercially viable. Small monochrome TN-LCDs with directly driven segments are used in digital watches, clocks, thermometers, measurement instruments, portable and static music systems, *etc*. Medium-sized TN-LCDs and STN-LCDs with multiplex addressing are used in calculators, handheld computer games, point-of-sale registers and mobile telephones, to name but a few applications. Large-area LCDs are used primarily in portable instruments, such as camcorders, personal digital assistants, notebooks and laptop computers. The variety of instruments and applications incorporating LCDs is reflected in the fact that 1.5 billion LCDs worth over $13 billion were sold in 1999. The success of LCD technology can be attributed, amongst several other essential factors, to continual progress in the design and synthesis of new classes of stable nematic liquid crystals with an improved combination of physical property values for a particular type of LCD for a specific electro-optical device application.

1 Portable Computers

In the short to medium term the most important market segment for LCDs in terms of value, volume growth and profitability will still be laptop computers and notebooks. Eighteen million laptop computers with an LCD screen were sold in 1999. These screens are relatively large (12–14″) and bright (> 100 200 cd m^{-2}), with full colour (> 250 000) and high information content, *e.g.* 1024 × 768 × 3 (RGB) pixels at high contrast (> 100:1) with short response times ($\approx$ 50–100 ms) and low power consumption (> 10 W). The great majority (92%) are TFT-TN-LCDs with active matrix addressing provided by thin-film transistors, while the rest incorporated an STN-LCD with a high degree of multiplex addressing. The optical performance of TFT-TN-LCDs is markedly superior to that of STN-LCDs of comparable size in spite of continual improvements in STN-LCD performance. However, the power consumption of STN-LCDs is much higher (× 4) and response times significantly longer (× 3) than those values of a comparable TFT-TN-LCD with a much lower contrast, *e.g.* 25:1 compared to 100:1. The market share of TFT-TN-LCDs has risen inexorably as the unit cost has decreased and the size has increased. New nematic liquid crystals with a low viscosity and a high dielectric anisotropy would allow lower voltages to be used, *e.g.* 3.3 V instead of 4.0 V. This would extend the battery lifetime by lowering the power consumption. A lower viscosity would also shorten response times towards video-rate addressing times. This would reduce blurring and increase resolution. A higher resistivity of the nematic mixture would increase the holding ratio of the device with a resultant increase in contrast. Continued progress in developing improved reflective polarisers using chiral nematic films (plus quarter-wave plate), multi-layer sheets or anisotropically scattering films would improve the viewing angle and brightness considerably.

2 Desktop Computers

Larger versions of the active-matrix addressed screens used in laptop computers will become increasingly important for use in non-portable desktop computer monitors (12″–21″). LCDs are set to steadily erode the market share of CRTs due to the advantages of low power consumption, *i.e.* lower running costs, absence of flicker and emission, portability and, most importantly, much smaller footprint. The cost disadvantage of LCDs of this size and complexity compared to that of CRTs is still more than a factor of three, but the price differential is decreasing constantly and quickly. The size of commercial LCDs at a realistic price for monitors is also increasing. However, the cost of a CRT for the monitor market is still very low and has been more-or-less constant in real terms for several years. The performance of CRTs is also being continually improved. The cost of an LCD with active matrix addressing of comparable size has fluctuated somewhat from year to year, but on average it has fallen substantially in real terms over the last five years. The total computer monitor market is large, *i.e.* 100 million units in 1999 and expanding rapidly. However, the growth in the number of LCD monitors is three times greater than that of CRTs, although from a relatively modest level (>3 million units). Three competing types of LCD with active matrix addressing share the market for LCD monitors. Four times as many TFT-TN-LCDs are still sold as LCDs based on in-plane switching (TFT-IPS-LCD) or vertically aligned nematics (VAN-TFT-LCD) technology, both with thin-film transistor active matrix addressing. However, the market share of the latter two LCD types is increasing steeply at the cost of the former. A lower viscosity of the nematic mixture would clearly improve optical performance of all three display types by reducing response times towards video rate. The inability of LCDs to operate at real video-rate frame times is the greatest limitation of LCD technology at the present time, more than size, cost, low brightness, insufficient contrast and limited viewing angles at acceptable contrast.

3 Mobile Telephones

In 1999 over 200 million mobile telephones were manufactured with an STN-LCD display unit. Nematic components with a larger dielectric anisotropy would allow lower drive voltages to be used. This would in turn reduce the power consumption and prolong the lifetime of the battery. A steeper electro-optical response curve would allow more information to be displayed. Monochrome OLEDs are an attractive option for mobile telephones from a consumer appeal aspect due to the ability for product differentiation and novelty.[5-12] Hybrid OLED-LCDs with one polariser replaced by an electroluminescent organic material emitting plane polarised light could also break into this market segment. PolyLED, a business unit of the Dutch consumer electronics division of Philips, intends to launch a monochrome OLED backlight using light-emitting polymers in 2001. Other contenders are surface-stabilised, chiral-nematic displays or bistable nematic displays, although these are still in the research and development stage. Larger LCDs with a higher information content are required for mobile telephones with internet capabilities.

4 Televisions

The very low cost and high performance of CRTs with small to very large screens (12″–42″) means that alternative technologies have failed to break into the market for non-portable TVs. The relative merits of CRT and LCD technology are reflected in the high market share of the former and the small market share of the latter for non-portable TVs and the inverse situation in the market for small, battery-driven, portable TVs. Very large plasma flat panels may capture a greater market share as their price continues to fall. The large number of processing steps during manufacture lead to the high production cost, rather than any other factors such as materials cost. However, lifetime, luminous efficiency and image quality are still issues to be addressed. Another possibility is a hybrid plasma-addressed LCD, where the plasma screen serves as a replacement for active addressing. These are potentially large-area screens with video-rate response times, full colour and high brightness. However, the practical problems involved and the cost of fabricating large LCD screens above 14″ is inhibiting the development of plasma addressed LCDs. Projection displays using small LCD screens (1″) with a very high pixel density have enjoyed some success, although the higher intrinsic brightness of digital micromirror displays (DMDs) suggests that they may occupy this niche in the TV market.

5 Miscellaneous Consumer Products

A surprisingly large number of consumer appliances incorporate an LCD, *e.g.* 1 million personal digital assistants, 4 million digital cameras and 7 million video cameras were produced in 1999. The requirement for short response times, low operating voltages, low power consumption and high information content with full colour of these devices means that competing flat panel displays will find it very difficult to enter this part of the market. However, OLEDs could well find application in car navigation systems due to the extreme temperature range specifications and relatively low and unsophisticated information content. Pioneer has already launched a range of monochrome OLEDs using small organic molecules for automotive applications.

6 Future Prospects

The major limitations and unsatisfactory aspects of LCDs are generally low brightness and restricted viewing angles with acceptable contrast.[1-14] These issues will have to be addressed in order for LCDs to maintain their position in the marketplace. More efficient backlights will certainly play a role in reducing power consumption and increasing brightness in standard LCDs. New types of LCD under development, which reflect incident light rather than transmit light from a backlight incorporated in the instrument, may offer a significant improvement in display brightness. Reflective and transflective LCDs with one or no absorptive polarisers may contribute to resolving these problems. Both reflective and transflective LCDs consume less power due to the use of reflected incident light. The absence of a backlight in a reflective LCD would lead to thinner, lighter displays with much lower power consumption. However, only a low proportion of incident light is reflected. A more promising kind of LCD is a black guest–host LCD with colour filters and active matrix addressing. However, the contrast is too low for practical applications. Therefore, the most promising approach is the further improvement in the physical properties of the components of nematic mixtures. Sub-pixellisation improves the viewing angle dependence of LCDs with the optical axis not in the plane of the cell, *e.g.* TN-LCDs and STN-LCDs. Therefore, this may be introduced in the medium term. This may require the introduction of non-contact alignment layers where the alignment direction is determined by the use of plane polarised UV light. Compensation layers will play a significant part in improving the optical performance of LCDs. Significantly increased size at acceptable cost is a much more intractable problem. However, this is a fabrication issue, rather than being dependent on material development. The use of plastic substrates will reduce weight and perhaps improve ruggedness for portable applications.

Of the emergent, competing flat-panel technologies large-area plasma display panels (PDPs) occupy a niche for very large direct view TV and as commercial, very large visual display units. This may expand into the consumer TV market with decreasing unit cost and improved resolution. Digital micromirror devices (DMDs) have great potential in projection displays for TVs and cinemas. However, along with flat-panel cathode ray tubes (CRTs) and vacuum fluorescence displays (VFDs), their considerable weight, high operating voltages and power consumption excludes this display type from portable applications. Interest in field emission displays (FEDs) has already waned and research and development is limited to a few specialist companies and universities.

Only organic light emitting diodes (OLEDs) using either organic low-molar-mass materials deposited by vacuum sublimation or polymers deposited by spin-coating or doctor blade techniques exhibit a combination of properties suitable for portable applications. The low-weight, power consumption and operating voltages combined with high brightness and good viewing angle dependence of contrast means that OLEDs are an attractive alternative technology to LCDs. However, the issue of power consumption and high information content will have to be successfully addressed, if they are to gain a significant market share of the flat panel displays market. High-information-content OLEDs with active matrix addressing exhibit twice the power consumption of a comparable LCD with backlighting. This is a direct consequence of their different modes of operation. OLEDs emit light on the passage of an electrical current and, therefore, current drivers are required. LCDs operate using a field effect where the optic axis is reoriented by the application of an electric field and very little power is required. The high power consumption of LCDs with backlighting is due to the backlight itself. This explains the interest in developing reflective or transflective LCDs.

The first production lines for OLEDs using low-molar-mass small molecules and conjugated polymers have been commissioned and products are appearing on the flat panel displays market. Pioneer is already supplying OLEDs (64 × 256 pixels) for automotive audio applications using technology licenced from Kodak involving low-molar-mass materials with red, green and blue elements. Sanyo and Kodak are collaborating on the development of a full-colour, high-information-content OLED panel (852 × 222 pixels) with active matrix addressing on polysilicon, rather than amorphous silicon, also using low-molar-mass electroluminescent molecules. A commercial product designed for camcorders and PDAs is expected in 2001. Philips has commissioned a production line in Heerlen in the Netherlands for the fabrication of OLEDs utilising light-emitting polymers. The initial use of these OLEDs produced under licence from CDT in the UK will probably be as backlights for LCDs. The Philips PolyLED business unit has also demonstrated a multiplexed OLED with more than 7000 pixels and 256 grey levels. CDT, initially a spin-off from the University of Cambridge, is a quoted company and currently has more than 50 employees. The UNIAX Corporation of Santa Barbara, USA, employs over 30 people and has also recently commissioned an OLED production line using light-emitting polymers. The first products include small (2″ diagonal) multiplexed OLEDs with over 600 pixels. Light-emitting polymers are manufactured by Covion Organic Semiconductors. Some of therse are manufactured under licence from CDT. A collaboration between CDT and Seiko-Epson has generated a full-colour OLED using ink-jet printing of soluble LEPS onto a poly-silicon active matrix substrate. OPSYS is a spin-off from Oxford University and employs more than 25 research staff. The prime interest of OPSYS is the development of electroluminescent metal chelates based on the lanthanides. This is also the case for ELAM-T, a start-up company linked to the South Bank University.

Therefore, although the market share of OLEDs is now very small, *i.e.* \$3 million in 1999, it is projected to increase significantly in the first years of the new millennium, *i.e.* \$500 million in 2005.[5] This will probably be at the expense of standard LEDs using inorganic semiconductors and vacuum fluorescent displays (VFDs), whose production volumes may actually decrease even though the overall market for flat panel displays is still expanding rapidly.[5] In the near term OLEDs will probably be used as backlights for LCDs and simple monochrome and colour displays with relatively low information content. Over the medium term OLEDs will probably be integrated in mobile telephones, PDAs camcorders and digital cameras. A potentially important market for OLED technology over the long term may well be light-weight FPDs supported on plastic substrates for a variety of portable devices. OLED technology on plastic substrates would have the added advantage of being compatible with low-cost roll-to-roll fabrication processes. Furthermore, Universal Display Corporation (UDC) of the USA has demonstrated flexible OLEDs on plastic substrates. OLEDs with a microcavity configuration offer the possibility of tuning the emission characteristics of bright OLED displays and, perhaps, the production of OLED lasers. However, the various research teams in companies and universities developing OLED technology will have to resolve a range of problems, such as device life-times, power efficiency, stable chromophores with matched chromaticity over time, patterning and addressing and, above all, large-scale, low-cost manufacture, before OLEDs are produced in bulk and start to capture significant market share from LCDs.

The significant growth in the number and value of all kinds of flat panel displays is set to continue. This is due, at least in part, to steady growth in the market for consumer products with digital displays and information-technology products such as mobile communications, calculators, personal digital assistants and portable computers. It is also partially due to the gradual erosion of the CRT share of the market for non-portable computer monitors. However, the explosive growth in the use of mobile telephones, especially with internet access in the future, is also making a very significant contribution to growth in the flat panel displays market despite the small size of the display. Mobile telephones with internet access will require larger, more sophisticated full-colour displays in the near future. A review of the materials chemistry aspects of LCDs and OLEDs, such as this monograph, suggests that new applications of both technologies are to be expected. This will be reflected in the volume and value of the high-value-added organic materials manufactured for use in LCDs and OLEDs. In turn this will help finance research and development in efforts to design and synthesise new liquid crystals and organic electroluminescent and charge-transport materials, both low-molar-mass materials and polymers, with an improved spectrum of physical properties.

7   References

1 D. Coates, E. Merck, UK, *personal communication.*
2 E. P. Raynes, *Proc. SID Euro Display '96,* 1996, 7.
3 T. Geelhaar, *Liq. Cryst.,* 1998, **24**, 91.
4 M. Thompson, *Proc. SID EID '98,* 1998, 6.1.

5 Stanford Resources, California, USA.
6 P. May, *Proc. SID Euro Display '96,* 1996, 613.
7 D. D. C. Bradley, *Proc. EMD Workshop '97,* 1997.
8 K. Pichler, *Proc. EMD Workshop '97,* 1997.
9 N. Bailey, *Information Display,* 2000, **3**, 12.
10 M. T. Johnson and A. Sempel, *Information Display,* 2000, **2**, 12.
11 L. J. Rothberg and A. J. Lovinger, *J. Mater. Res.,* 1996, **11**, 3174.
R. H. Friend, R. W. Gymer, A. B. Holmes, J. H. Burroughes, R. N. Marks, C. Talliani, D. D. C. Bradley, D. A. Dos Santos, J. L. Brédas, M. Lögdlund and W. R. Salaneck, *Nature,* 1999, **397**, 121.
13 J. R. Sheats and P. F. Barbara, *Acc. Chem. Res.,* 1999, **32**, 191.
14 S. Forrest, P. E. Burrows and M. E. Thompson, *Chem. Ind.,* 1998, **12**, 1023.

Subject Index